MOTHER'S DAY

THE COURAGE & SACRIFICE OF THE 3RD BATTALION 25TH MARINES

TAYLOR CLEVELAND

TACTICAL 16 PUBLISHING

MOTHER'S DAY

The events and conversations in this book have been set down to the best of the author's ability, although some names and details may have been changed to protect the privacy of individuals.

First Edition

Published by Tactical 16 Publishing

Colorado Springs, Colorado

www.Tactical16.com

ISBN: 978-1-943226-83-2 (paperback)

CONTENTS

THE SERGEANT MAJOR OF THE MARINE CORPS

4 August 2020

To the Marines and Sailors of 3rd Battalion, 25th Marines,

Fifteen years have passed since the warfighters of 3rd Battalion, 25th Marines responded to our Nation's call to fight the Iraqi insurgency. The Battalion's fight in Haditha was the single deadliest deployment for a Marine battalion since that of the Beirut Barracks bombing in 1983.

Today, we remember these Marines and Sailors who proudly self-sacrificed their personal commitments and went forward to push the cause of freedom. They were emblematic of the citizen soldiers who serve as beacons of the community by living the Marine Corps' story of daily service to family, community, and nation.

We honor the Lima Company Marines and Sailors that paid the ultimate sacrifice in the preservation of freedom for future generations and live lives worthy of their sacrifice. Mission First, Marines Always!

Semper Fidelis,

TROY E. BLACK
19th Sergeant Major of the Marine Corps

FOREWORD

From February to October of 2005, it was the greatest privilege of my life to serve with the thousand men and women of the 3rd Battalion, 25th Marine Regiment (3/25) in the Al-Anbar province of Iraq. Those eight months have shaped every minute of my life since. We trained as a unit, deployed as a unit, and fought as a unit. But even though we were one unit in one war, there are a thousand individual stories about that deployment. Mine is only one of them. In combat, I was limited to experiencing what happened in my immediate area, understanding it from my own perspective.

I am not an authority on Iraq or the 3/25th's role in that war. I can only tell my little piece. I tried my best to be accurate in the accounts of my experiences, but I completely understand that people who stood right next to me may remember things differently. Maybe they were privileged to have more information than I was that made them see a different side of the same event. Everyone's war is different.

This also is certainly not a comprehensive list of stories from the 3/25 needing to be told. For the eight months that we operated in Iraq, we were the most active combat unit in the country. At the end of our deployment, we had the highest casualty rate of any single battalion for the entirety of both wars in Iraq and Afghanistan. I know there are stories of heroism and sacrifice that could fill a hundred books. But I was only there for a fraction of them, and these are my accounts. For those stories that are not in this book, I apologize.

The names in this book are published with specific permission from each individual or their family; or where that person's name is within the public domain. There are important people in this book from whom I could not obtain permission; those names are pseudonyms. My intention is to be as honest as possible. However, everyone has an individual right to privacy, and I respect my friends' right to have their roles in these stories told on their terms.

CHAPTER 1
WHY

Growing up, I had a lot of veterans who played important roles in my life, my father included. I felt I had a good appreciation for their service, if not their sacrifice. Most of the veterans I knew were combat veterans, having fought in WWII, Korea, and Vietnam. I was always in awe of what they had been through. For me, those who fought in war were about as close to superheroes as they could get.

Back then, talking to any veteran about their experiences in war was unheard of, especially with my father. It was just something we didn't do. I don't know why, whether it was out of respect, or out of some misplaced fear that the mere mention of their war could dig up ghosts from their past that they had worked for years to bury.

In fact, the first time I ever talked to my father about his service in Vietnam was for a college project as part of a "History of the Vietnam War" class. I was twenty-nine years old and waiting to be deployed to Iraq myself.

The interview turned quickly from a scripted school project to an effortless father and son conversation. I vividly remember his emotions as he shared his tragic and comedic experiences. He laughed hysterically when he told me of the time he walked drunk into the back side of an elephant. His eyes filled with tears as he recounted curling up in a ball as rockets exploded around him. His story consumed me to the point that I found myself holding my breath to see what came next.

Interviewing my father for that project was the only time he ever talked to me about Vietnam. I remember at that time thinking my own impending deployment had probably earned me the right to listen to his story. Maybe it had. When I returned from Iraq, I found myself in a similar position, asked by some close friends what it had been like to fight in a war. When I talked to them, all guys who had not served in the military, it felt good. It made me feel like people cared.

When I thought back to my dad's interview, it made me think that my father's story was something he had wanted to tell. He was just waiting for the right time and all it took was someone who cared enough to ask.

While that interview made me feel closer to my father, it still seemed like I was an outsider looking into his world. It was like he wasn't able to tell the complete truth as he recalled it. He wasn't holding anything back or trying to protect me from the pain that he had experienced; he was just a man trying to explain extraordinary experiences in a way ordinary people would understand. As riveting as those depictions were, they told only part of the story and contained only a sliver of the emotion. That was as close as I had ever gotten in my early life to understanding what veterans went through, and it was vastly deeper than any conversation I had experienced up until that point in my life.

After I returned from Iraq, the Vietnam vets that I had had superficial conversations with throughout my youth began to seek me out and talked to me about my own deployment. These were men that had been my Boy Scout leaders, teachers, relatives, priests, and family friends; men whom I had known and whose service I had known about for my entire life. But after my own wartime experiences, things between us had changed. It was like they had been waiting for years to talk to me, or to at least someone like me. I was a different person now, and somehow, they were able to talk to the post Iraq me in a way they had not been able to before, in spite of our friendships. The look, to a man, on every one of their faces seemed to say what had been bottled up inside for forty years. "Now you know. Now you know the secrets we kept all this time. Now you know what service, and sacrifice mean."

Every time I hear a veteran's story, no matter what war they fought in, I find myself searching for the right thing to say, but the right thing never comes. All I am ever able to do is just listen patiently to what they have waited most of a lifetime to say. I think as I listen how terrible it must have been for them to have kept that part of them inside for all these years; and I am deeply humbled that they trusted me enough to share it.

I have been back from Iraq for over fifteen years, and I have never had a problem telling people about what I did or how I felt when I was there. And I certainly never found a shortage of people willing to listen. But the same chasm exists for me when I try to explain Iraq to people who are not veterans as it did for my father trying to explain Vietnam to me. I feel bad for them, my audience. It's not that they don't want to understand…it's just that I don't have the words; I don't even know if the right words exist. I mean, how in the hell do I explain a war?

Even though I do not have a problem talking about my time in Iraq, plenty of people that I served with do. I understand. Some people want to put the war in their rearview mirror. Some want to talk about it but don't know how or are scared of what will happen when they do.

I have a lot of sympathy for the warriors who want to tell their story but struggle to find the strength. I know what it's like to tell war stories to ordinary people and wonder, "am I going to keep my composure or am I going to break down crying at something as simple as remembering what a palm grove smells like?" I have wondered if I could even organize my thoughts into words that made sense to folks that may not even know the US still has troops in harm's way. After all, we are trying to explain the military to non-military people. There's a language and culture gap that's huge.

I know what it is like to worry that people will judge me; to be afraid that people will think that I feel less of a man because of things that I did, or things that I wasn't strong enough to do.

I understand the balance between being honest and being brutally honest. In war, truth is more complex and graphic than anyone can possibly imagine. I know the tendency when telling a story to hold back or gloss over the really bad parts, because I just am not sure that ordinary people can handle it.

Most of all, I understand how difficult it is to talk about oneself. I served with some of the most selfless human beings to ever walk the earth. Men who literally died trying to save other men they may have only known for a few months. But when someone talks about war, it is from their perspective. It is inherently self-centered and deals mostly with the experiences that effected that particular individual. Invariably, when anyone tells a war story, they bump into one or several instances where they acted in a way that an objective person would describe as "heroic." No living person likes to be called a hero. That's a word that well-meaning people throw around to express their appreciation; but it just ends up making the guy on the receiving end feel empty or ashamed. No one wants to talk about when they acted "heroically" and their friends still died.

No matter how difficult it is, I think that there is immeasurable value in every veteran telling his or her story to the degree that they are able. I think of all of the lost experiences as the last WWII veterans pass away in silence. I worry that this is now starting to happen to our Korean and Vietnam veterans. I truly believe people want to hear our stories. And I truly believe that those of us who can tell our story have an obligation to do so; if for no other reason than to let the world know about the magnificent human beings that either gave their lives in service of their country or returned injured in a way that keeps them from being heard.

I am writing with the hope that other veterans, no matter when or how they served, will understand the value that their individual experiences have to the nation and that they will choose to share them. I am, of course, also writing for those who no longer can write for themselves. I will hold their memories in my heart forever and I sincerely hope that our country will do the same.

So that's why.

CHAPTER 2
THE CRASH

Shortly after two in the morning, the tones dropped. A loud, deep, piercing sound came from a pager sitting on my nightstand. The sound lasted just long enough to awaken me to the point where I was able to hear the words that followed. "Station #1 for an Engine Company, Station #3 for Squad and Rescue. 442 West Main Road for a motor vehicle crash with entrapment. One vehicle into a pole. Caller reports multiple victims trapped."

That night, just like thousands of nights before, I was working as an on-call paramedic for the Conneaut City Fire Department. It was a job that I had held since leaving high school ten years before. That job had helped me work my way through college. And even though I had graduated and was working full time as a Detective in the Police Department for the next city over, I stayed with the Fire Department. Even though I was weeks away from deploying to Iraq with my Marine Corps Reserve unit, I continued to pick up shifts. I liked the work. It was important. I made a difference. Most days at least.

The tones set off by the dispatcher started a carefully coordinated response. A fire engine staffed at Station #1 by the city's full time firefighters, pulled out of their station about half a minute after the dispatch. At Station #3, two part-time EMTs staffing one of the city's ambulances, or "squads," would leave their station about the same time. A few minutes later, another crew of part-time firefighters would arrive at Station #3, prepared with the Heavy Rescue, a truck filled with the technical rescue equipment needed to cut the car apart and free the trapped patients.

Since I lived about twelve minutes from my station, Station #3, I kept my gear on the squad and thus available to drive directly to the scene to assist the EMTs with a higher level of medical care, should they need it. And that night, it sounded like they would.

With the tones still fresh in my ears, I swung my legs over the side of my bed and sat there for about ten seconds trying to process what I had just heard. Crash. Entrapment. Victims. Pole. Each one of those words held a particular significance. Each word meant a different decision to make, different equipment that would be needed, a different contingency to plan for, but having been awake for less than ten seconds, I didn't yet have the cognitive capacity to make them. I focused first on sliding into my coveralls, grabbing my car keys and portable radio, and making my way out the front door of my house to my pickup truck. As I moved out of my bedroom and down the stairs, my wife lifted her head and asked quietly, "Honey, please be careful."

And like a thousand nights before, I assured her, "I always am."

As I hopped into my Toyota pickup, the words that I heard during the dispatch came back to me. I sorted out what each one meant in my head and started coming up with a plan.

Crash. Car crashes during the day were often insignificant. A result of someone not paying attention and rear ending another car. The injuries were often minor and uncomplicated. Car crashes at night were often alcohol related. The speeds were much higher. The impacts are much more violent. The injuries are vastly more complex. Patients involved in these crashes needed to be in a level one trauma center (a hospital with a full spectrum of neuro, cardiothoracic, vascular and trauma surgeons on staff) within one hour of the crash if they were to survive. Our local hospital was only a small receiving facility with less than twenty beds and only an emergency room physician working. Handling complex trauma patients from a motor vehicle crash was well beyond its scope of care. The nearest trauma center was a half hour away by helicopter. As I threw my truck in drive, I took note of the weather, making sure it was clear enough for the medical helicopter to fly. I then started mentally working through a series of potential landing zones near the crash site. I radioed dispatch and asked that they put a medical helicopter on standby. A thirty-minute flight time meant that we had precious few minutes to spare, and I thought that giving the helicopter crew a couple extra minutes to prepare their aircraft and shake out the cobwebs might be a lifesaver.

Entrapment. Modern cars are built to impeccable safety standards. Side curtain airbags that encase you in a bubble of air filled pillows, impact crush zones that distribute the force of the crash away from the passenger compartment, and seatbelts with pyrotechnic pretensioners that automatically tighten up upon impact to keep you from moving in your seat. But those safety standards only went so far. Enough force can tear a car in two; or wrap it like a pretzel around a pole; or crash it head on into another car so that both of them become indistinguishable as individual vehicles and merge into one giant metal tomb. If a crash generated enough force to disfigure a vehicle in such a way that the occupants were trapped inside, then the only way to get them out was by cutting what used to be the car apart, piece by piece. The heavy hydraulic equipment (called a Holmatro) was in the Rescue Truck. That Rescue Truck would be leaving Station #3 about three minutes after the crash. It would take it about seven minutes to get to the crash scene. It would take another five minutes to set up the equipment and about twenty minutes to cut out the patients. Then another ten minutes to treat, package, and transport the patients to the landing zone. Plus, the thirty-minute flight time. That put us one hour and fifteen minutes from the time of the crash to arrival at a level one trauma center. Not acceptable. Somewhere, I needed to shave off fifteen minutes if the patients were to have a fighting chance.

Victims. Plural. More than one. How many more? No clue. If there were too many, I would need to call another ambulance from another city. That would be another fifteen minutes at least. That could be a death sentence. Options started scrolling through my head. I could treat two patients in one ambulance. That would save time, but I would have to divide my attention and treat each patient with half the staff, and half the equipment, that I would normally have at my disposal. I only had one heart monitor, who would get it? Can one EMT manually breathe for two patients by squeezing a ventilation bag with each hand? How long could I do that? What if there were three, or four patients; who do I treat first? Who do I treat last? Who would I not treat at all because I didn't have the equipment or the manpower? Who would die because of what I couldn't do?

Pole. Cars into poles meant the potential for downed electrical wires. If those wires were down, then the electric company had to be called to shut the power off to enable Fire and EMS to extricate and treat the patients without themselves being electrocuted. The electric company's emergency response was usually between forty-five minutes to an hour...a timeframe that would not be survivable by anyone. I said a quick prayer, "Please God, don't let there be power lines." I did not have a plan for that.

As it happened, the crash site was about a mile away from my house, and about six miles away from the closest fire station. I would be arriving at the crash a full five minutes before the first engine and squad; and all my equipment was on that squad.

As I rounded the corner in my Toyota, I could see the smoke from the smoldering car fifteen seconds before I saw the crash. A sedan had failed to round a curve in the four lane road and struck the pole on the apex of the curve at an incredible speed. The speed limit on the road was fifty-five miles per hour, but no one ever went that slow. God only knows how fast that they were going, but the car they were in was barely recognizable as a motor vehicle.

The car was upside down, on its roof, and the passenger side was partly wrapped around the telephone pole. A small crowd of bystanders from a nearby apartment complex had gathered about fifty yards away from the crash and were pointing and screaming. No one was trying to help the people inside. The car was smoking but it didn't appear to be on fire. As I was trying to process what I was seeing, I silently cursed the gawkers for not helping.

I parked my truck and ran to the scene, through the semi-circle of onlookers. Once I broke through their ranks, I stopped dead in my tracks. I understood now why none of the civilians would go any closer: the car was smoking because there were live power lines on top of it.

Balls. Of all the lousy luck. I took a deep breath, stepped back, looked around and tried to figure out what this meant. As I traced the power lines back up the pole, I could see that they were primary power lines. Power lines come in two types. Secondary lines bring electricity from the pole to a house. They have a lower voltage; still enough to kill if touched, but there are some options to work with to make them safe. Primary lines carry electricity to whole city blocks. Not only can a touch kill with these, but they can also "jump" and complete a circuit to kill anyone too close to them. So, there were wires down on the car, and of course, they were the bad kind.

I moved as close to the car as I could to try and get a rough assessment of the victims. How many were there? Were they still alive? What were their injuries? How complicated would the extrication be? As I approached the driver's side, I noticed that the driver's door was pushed in about a foot from where it would normally be. The driver, who looked to be a male, was horribly disfigured. I stared at his chest for a full thirty seconds. Looking for any movement at all that would suggest that he was breathing. There was none. I wasn't sure he was dead, but I knew that if he wasn't, there was nothing that I could do for his massive systemic injuries.

I moved from the driver's side of the car to the next problem at the passenger side. What used to be the passenger window was almost entirely obscured by the telephone pole that was now inside of it. I laid down on the ground next to the pole and was able to see the passenger compartment through a void about the size of my fist. There, inside of the car, was a girl. Not very old, maybe nineteen or twenty. My sister's age. She was suspended upside down by her seatbelt. Her head had struck the telephone pole that was in her window, causing a massive head injury. She was breathing but was struggling to do so. Each breath progressively got slower and shallower. She was dying. Her head and face were swelling from the trauma. In a couple of minutes, her airway would close, her lungs would fight unsuccessfully a few final times to pull air past the blocked airway, and then she would die.

I popped up from the side of the car, grabbed my radio and gave my assessment of the scene to the dispatcher. "EC-21 to Conneaut Fire, on location. One vehicle into a pole with entrapment. Two priority one patients. Power lines down on the car."

Although my assessment didn't give the other responding units a lot more information than we had during the initial call, everyone now knew for certain what we were dealing with on scene. Any question of how serious or complicated this would be, was now out of everyone's mind.

A few short moments later, the engine company pulled up. I walked up to the lieutenant in charge and gave him my "no shit" assessment. "LT, two kids in the car, I think the driver is dead but I'm not certain. The passenger is working on dying and she is going to get there real fast if I don't get her intubated." (Intubation is a procedure where a plastic breathing tube is placed into the mouth of a patient, into their trachea. A bag is attached to the tube allowing someone to squeeze the bag and manually force air into the patient's lungs, breathing for her when she cannot breathe for herself.)

As the lieutenant spoke to me, I heard fear and exasperation in his voice. "The electric company is on the way. Forty-five minutes out." Next to the lieutenant were the other two firefighters from the engine. None of the three of them were moving. It was obvious that the plan was to wait for the electric company to move the lines before we did anything. We had just sentenced a young woman to death.

All I could see in my mind at that moment was my little sister, hanging upside down in that car, taking her last breaths, with no one there to even hold her hand, or lie to her and tell her it was going to be okay. And next to her, I saw myself. Standing there. Doing nothing. Watching her die.

It took about a second for me to decide that was unacceptable. I yelled to one of the firefighters from the engine. "Chris, grab the

TAC stick and let's get this thing going." He didn't move. "Come on man, let's get the TAC stick and get the fuck on with it!" A TAC stick was a three-foot-long electrical meter that we used to determine if a downed wire was energized or not energized. If I could get close enough to the primary lines with the TAC stick and determine that they were not energized, we would be in business. If they were energized...then that would have been a mistake.

Chris grabbed the TAC stick from the rig and handed it to me. I turned it on and inched closer to the line, foolishly thinking I could jump away once I started to be electrocuted. I checked the first line. Dead. The second one was dead as well. One by one I confirmed that the lines on the car were not energized. After the last one, I ran back to the crew of the engine. "Were good! Let's get at it."

Still no movement from the other firefighters.

They knew we weren't "good."

When a power line comes off of a pole in a case like this, most of the time it automatically trips a transformer. That shuts off the power to the line, making it safe. Which was exactly what happened in this case. However, some of those transformers have a timed reset feature, meaning that they will re-energize every so often just in case whatever caused them to trip in the first place was a temporary event, like the wind blowing two power lines together. So even though the power lines on that car were dead now, at some point in time a surge of power was going to come rushing back through them for a brief second. Anyone caught in, on, or near that car when that happened, was going to get electrocuted. How long did we have until that happened? I had no clue. It depended on the transformer. Ten minutes, twenty minutes maybe. And it had already been about eight minutes since the crash.

I could tell from the looks on everyone's face that this was a risk they were not willing to take. We would wait. She would die.

I don't even remember making a conscious decision to move. The next thing I remember was crawling into the driver's side of the car. The driver's window had been made into a hole just large enough for me to snake my way into the car. As I crawled inside, I crawled onto the driver and started feeling his body, starting at his head, for signs of life. The inside of the car was completely black. I could not see him, or the girl, at all. All of my assessment and treatment would have to be done by feel. As I ran my hands across his head, I could feel how violently it had been struck, probably by the pole. I placed my hands in front of where his mouth and nose had been, and put my ear against his chest, hoping to hear or feel any sign of life. I laid there, on top of him, and counted to fifteen. Nothing. He was gone.

I crawled further across the dead boy, our bodies tangled together like I was laying in his lap, and he was hugging me. I could hear the sound of the girl, unconscious, but whimpering as she fought for each breath. I reached towards where the sound was coming from and pulled open her mouth, clearing a path for her to breath just slightly better.

Just then I heard Chris. His face was at the hole in the passenger side of the car, near the bottom of the pole, where I had been earlier. "What the fuck are you doing, man?" he yelled at me.

"Chris, what's the ETA on the helicopter?"

"Fifteen minutes," he shot back.

"Tell them to land at the hospital, we will meet them there and do a hot offload on the pad." The helicopter was already enroute to our location but landing them at the scene was too dangerous. There were only a few ways this scene could get worse, and a helicopter causing flying electrical wires, and potentially a helicopter crash, was definitely one of them. The hospital had a solid landing site and was only a couple of minutes from the crash scene. It was the best option.

Chris didn't seem sure of that plan. "You think we're getting her out of this car in fifteen minutes?" he asked, making sure that I knew that what I was suggesting was not possible.

"Yeah, Chris, we are getting her out, and we are getting her out alive."

I could hear Chris being still for a moment, letting the whole plan sink in, trying to figure out the move. In order for this to work, they would have to cut the car apart in record time. Not only to shave off the fifteen minutes needed to get the girl to a trauma center in time, but to make sure that none of us would be electrocuted when the transformer reset. Then, after a moment of silence, I heard him take a breath and start yelling orders to the squad and Heavy Rescue that had just arrived. We were in business.

I could hear the crew of the Heavy Rescue, pounding wooden blocks in around the car to stabilize it, and starting the generators on the hydraulic Holmatro which would cut the car apart. Chris kneeled back down beside the pole and yelled into me, asking if there was anything I needed. "Give me a number seven ET tube and bag valve mask. No need for the scope, I can't use it in here."

An ET, or endotracheal tube, is the breathing tube inserted during intubation. I had felt the girl's jawline with my hand and estimated the size tube I would need. If I successfully inserted the tube, I could use the bag to breathe for her and buy her some time until she reached the trauma center. Her breathing had slowed considerably from when I first saw her, a result of her airway swelling. I didn't know if I was too late.

To place an ET tube down someone's windpipe, we use a tool called a laryngoscope. It's meant to be used in a very specific way. Normally, the patient lies on their back, and we approach the patient's head from the top, oftentimes lying on our stomachs to do so. The scope then goes into their mouth, and a curved shape "blade" goes down past the uvula (the hangy downy thingy in the back of your throat). We then pull up on the scope and the uvula is moved away, clearing the way for us to see the windpipe and insert the tube where it needs to go.

I could not do any of this. It was pitch black so I could not see. I could not move my body and could only move my hands inches, and the girl was hanging upside down from a seatbelt. Even under normal circumstances, paramedics screw intubations up a little more than half the time. This was anything but normal.

As Chris passed me the ET tube through the hole, I scooted my face closer to the girl's face, so that my nose was almost pressed against her cheek. I wanted to take advantage of everything I could hear or feel when I was trying this. The two fingers on my left hand were already holding her mouth open to breathe. I slid the tube into her mouth and held it tight with my left thumb. I then reached three fingers on my right hand as far back into the girl's mouth as I could, moving her uvula out of the way with my index finger and feeling her epiglottis (the flap that covers your windpipe) with my middle finger. I used my left hand to pass the tube further into the girl's mouth, to where my fingers on my right hand could grasp it and guide it into her trachea. I felt the tube slide surely into place and secured it by inflating a small cuff with a syringe full of air.

I then hooked the bag valve mask up to the end of the tube and started squeezing it to breath for her, counting six seconds in between each breath. I could feel with my left hand her chest fill with air each time I squeezed the bag.

I yelled to Chris that I had secured an airway and asked him how it was going with the extrication. I could feel the heavy Holmatro tools starting to pry apart the car. "Not good man, they are having a hard time cutting around you. You're gonna need to come out. We're going to take that driver's door off, pull out the driver and then pull out the passenger. We can't get to her from my side because the pole is in the way."

I took a deep breath and told Chris in the calmest tone I could, "I can't do that man. I'm not getting out of this car."

Chris yelled back, irate, "Fuck you man, don't be a fucking hero. Let me do my fucking job and get out of that car!"

I tried to slow my breathing down and explain my situation to Chris. There were tears in my eyes as I told him barely above a whisper, "Chris, I'm not being a hero. I can't get out. I'm trapped in here. You have to cut me out, too."

The hole I crawled in through at the driver's window looked like the hole made in the side of a Pepsi can when it's shot it with a gun. The metal mushrooms out on the inside. If someone put their finger in the hole of that Pepsi can, it would slide in easily. But if they tried to pull it out, the jagged pieces of metal would tear at the finger and trap it inside. I had made it into the car, but there was no way I could get out.

I laid in that car and listened to Chris and the Heavy Rescue crew figure out what to do now. Their whole plan had been to start cutting where I was, once I got out. Now that I, too, was trapped;

that plan was impossible. For the next minute, I laid there. On top of the driver's dead body. With my face pressed against that girl's cheek. Breathing for her and whispering to her that it was going to be okay. Listening to the firefighters dismiss one new plan after another as unworkable.

I tried to think of a plan. I tried to figure out what to do. That's when I felt it. It was cool, cold almost, and it started running down the back of my neck like someone was running a hose down my shirt at a trickle. I was instantly sure what it was. Blood, from the girl. She was bleeding out. She had a hemorrhage that I couldn't see and couldn't treat, and now she was going to bleed to death as I lay next to her in that car. For about ten seconds, I tried to figure out how to locate the bleeding and options to stop it. All the while the liquid was flowing down my shoulder, to my neck, and covering my face.

It ran into my mouth and there was so much I had to spit it out to be able to breath. That's when I tasted it. It wasn't blood.

It was gasoline.

The car had been leaking fuel and my clothes...and the girl's clothes were now saturated in gasoline. It was pouring out of the tank and covering our bodies. Before that moment I couldn't think of what could be worse than being electrocuted while you were trapped in a car.

But this was worse. Being electrocuted, trapped in a car and blowing up in a flaming ball of gasoline. That was worse. Hands down.

I raised my voice enough so Chris could hear, "Chris? Hey buddy, the car's leaking gas. A lot. We're covered in it. We don't have much time, man. I need you to hurry...Please." The "please" wasn't me being polite. It was me asking him to save my life and the girl's next to me. Chris knew exactly what that meant and he wasted no time snapping to.

Immediately after that, Chris and the Heavy Rescue crew tore into where the passenger side back door on the car used to be, just to the side of the telephone pole. They cut a hole through the metal and removed the back seat, behind the girl. Then they started cutting apart the girl's seat, gently, with her in it.

The girl was the first to come out. As light poured in though the fresh cut opening in the side of the car, I could see her entire body for the first time. It was broken. Badly. She was pulled out and put immediately on a stretcher. As soon as I gained a few inches, I jumped out of the car and started working on the girl as we moved to the back of the ambulance. Heart monitors, IVs, splints, and bandages were all applied in the back of the squad as we raced the two miles to the hospital. The girl was clinging to life. Barely.

We arrived at the hospital as the helicopter touched down. We pulled the girl out of the ambulance, on the stretcher, and wheeled her towards the crew of the medical helicopter who met us in a field halfway between our truck and their aircraft. We ducked low to avoid the rotor wash as we transferred the girl into their care. I gave a brief medical report to the flight nurse about what had happened and what I had done. Within two minutes, the helicopter lifted off on their thirty-minute journey to the trauma center: and she was alive.

It had been about twenty-six minutes since the crash, putting her at the hospital in under an hour.

As I write this now, it seems foolish. But at the time, I was elated. I might have even cheered as the helicopter took off. We as a team had risked our lives and done something that seemed impossible. We had made a difference. We had saved a life.

When the girl died, I was devastated. She had gone to school with my sister. They were friends. She was a bright, vibrant young woman, and the world would have been a better place with her in it. Her boyfriend had been driving. He also was a fine young man. Everything we had risked was for nothing. I had not made a difference. No lives were saved.

That night, as I ran toward the squad with the girl, I brushed past Chris. Not a word was said. I was focused on the task at hand, but I never told him thank you. I never told him how much I appreciated him saving my life. Chris Horwood would go on to spend a year as a contract firefighter in Iraq. When he returned, he suffered from a debilitating illness that was most likely linked to his exposure to burning chemicals while overseas. Chris died a few years later, still not having received a thank you from me. Chris Horwood was a good man, and I am forever in his debt.

A short time later I would deploy with the reserves to the Al-Anbar province of Iraq as a medic for a Marine Corps Infantry Unit. The skills that I had learned in my medical and law enforcement careers over the last ten years would be put to use daily.

But just as important as my medical training would be the lessons that I had learned in a decade of trying to piece broken people back together. People get hurt. Badly. And some people that I count on wouldn't risk their lives to help them. That meant that sometimes I was going to have to risk mine to help them. But I also knew that, sometimes, people would risk their lives to save me. I knew that there was being brave, and there was being reckless and sometimes those things were as far apart as heaven and hell. And some people spend their whole lives trying to figure out which was which. And I knew that sometimes, no matter what I did, people would still die. Better people than me. People who would have made this world a better place if they had lived.

CHAPTER 3
IN THE BEGINNING

When I returned from Iraq, I was talking to a friend of mine who was a Vietnam veteran and former police officer about his war. His description of the impact that his service had on him sticks with me to this day.

"I don't even remember who I was before the war. It's like my whole life began over there, that's where I found out who I was."

He was nineteen when he went to Vietnam. His whole adult perspective for his entire life was formed for him in thirteen months. I think about the thousand Marines of my battalion, and how many young boys were born into adulthood from war.

I remember exactly who I was before the war. I was a father, a son, a husband, a police officer, and an American. That was the same person that I am now…with just a little more perspective.

It's surprising to me that only after having become a veteran, could I truly appreciate how much of an impact that veterans, the values they hold, and the lessons they taught, had shaped my life.

I had dressed up in camouflage every year for Halloween from the ages of six to fourteen, for I had been playing Army with the kids in my neighborhood since I was in kindergarten. Most of us were way too poor to actually own a toy gun, so we used sticks or tied pieces of pipe together. Every once in a while, some kid's dad would use a jigsaw to make a gun out of piece of scrap wood and that kid would be so admired that he would be immediately promoted to squad leader. For my friends, we gauged how much money a kid's family had by what kind of fake gun he had. A stick that found in the woods, dirt poor; a piece of pipe used for a bazooka, medium poor; gun carved out of scrap wood meant their dad worked at a good job in a shop with tools and time for side projects; actual store-bought gun, rich.

We outfitted our makeshift army with fatigues that my uncle had left over from Vietnam. I kept two sea bags in a fort that my dad made out of an old cable reel he got from work at the phone company. We could only fit about four kids in it, but it made a pretty good base of operations.

We would have about eight to ten kids at any one time. Sometimes we would pick teams and fight against each other, sometimes we would fight against hordes of imaginary enemies. Whoever we fought, there was always a twenty-five second rule that would bring you back to life.

My father was a Marine in Vietnam who served with the 2nd Marine Air Wing in Da-Nang. My uncle, Andy, was an Army Ranger and platoon leader who was wounded severely in Vietnam and received a Bronze Star. My grandfather, Wilson, was in the Navy during WWII and had two destroyers sunk out from under him; the first stranded him on a desert island in the Pacific for a month and the second during the invasion of France on D-Day. My grandfather Brown was wounded in the Battle of the Bulge as a tanker with an Army armored division. My great grandfather, Taylor, was in his fifties, when he fought in the Philippines during WWII and my great grandfather, Wilson, served as a medic in WWI. My family can trace its military lineage back to my nine times great grandfather, Moses Cleveland, who served with George Washington in 1776. For me, military service was kind of the family business.

My Boy Scout leaders were Korea and Vietnam veterans, as were my high school teachers. Even my priest had served in WWII. Everyone in my small Ohio town had either served in the military or had a relative who had. It was sort of understood that in Conneaut, Ohio, that was just what we did.

A result of that near universal community service was a unity of values that were passed on from one generation to the next. We worked hard, went to church, served our country, volunteered at the fire department, voted, planted trees, and hunted deer. It was a simple set of values centered on community and family. It was good and wholesome and wonderful, and I completely took it for granted until I returned in October of 2005.

Everyone remembers where they were on September 11th, 2001. They remember what they were doing, how they felt, and who they talked to. That moment in history will forever be filed away in the brain of every American of a certain age, next to their first kiss, their sixteenth birthday, and their wedding day. It's just a picture that you always keep in your head.

I was driving home from Kent State University. The sky was blue...so blue, the temperature was perfect, just a little breeze. The leaves hadn't started to change colors yet, but they weren't as green as they were in August. I had just finished class for the day and was going home to change for work. I was working the swing shift that day and I was going to have just enough time to eat lunch and get dressed before I went in to work.

September 11th came well into my life. I was married; I owned a home, my wife, Louise, was expecting our first child, and I had just been promoted at the police department from patrolman to detective. Life was good for me. I remember hearing the first reports. A plane had struck the World Trade Center. Like most, I thought that it was an accident, maybe someone wasn't paying attention and flew a Cessna into the side of it. Then, the second plane hit. The news dominated the radio. It's hard to imagine now, but in 2001, only flip phones existed, and we still paid for minutes...to talk. Texting wasn't a thing. I pulled into my driveway and turned on the TV. News had taken over every channel. Video of the towers burning made me sick to my stomach.

At work, for training, we listened to the 911 and radio calls from the first bombing at the World Trade Center in 1993. That had been a mess. That had been the biggest disaster I had ever seen an agency work through, but this was going to be a thousand times worse. Even after that second plane hit, I still didn't think that we were under attack. I thought it was just a terrible accident.

Then the third plane crashed into the Pentagon. There was no doubt now; we, me, my country, my town...were under attack. I wasn't sure by whom, but I was certain that this was the first shot fired in what was to be a long war.

I was scared. I was scared for my country, my family, and my future. I don't ever remember feeling such an abstract fear, such helplessness from knowing that there were well organized acts of violence being perpetrated by people with the sole intention of killing Americans.

I imagined this was the way my Grandfather Wilson had felt during the attack on Pearl Harbor, and the day after, when he marched down to the recruiting office to join the Navy.

For the first time in my life, I also felt ashamed. Ashamed of the choices I had made in my life. I was ashamed of one key life choice, really; the choice to not serve in the military.

I had always planned on joining the Marines directly out of high school. There never was any other plan. My parents had no money for college. It was just what everyone had assumed I would do.

In 1993, my junior year in high school, I visited a recruiter and started what I thought was my journey into the Marine Corps. The first stop was the required medical examination at the Military Entrance and Processing Center (MEPS) with about a hundred other skinny teenagers who had the same ambition I did. I remember sitting down with the doctor and answering a litany of medical screening questions. Everything was going fine, I thought, until he asked if I had had any surgeries. "Sure" I said, "on my knee". The doctor, who had been previously very disinterested in my answers, my presence, or even my existence, suddenly became very interested in my knee.

"What kind of surgery did you have?" The words came out of his mouth so slowly and deliberately it was almost unnatural. It gave me the impression that the answer to this next question could determine the fate of the rest of my life.

"An ACL reconstruction."

I played football during my freshman year in high school. Although I wasn't blessed with any real athletic ability, I could run, and I could tackle people. While I wasn't good enough to start, I was for sure going to letter my freshman year. That was until my legs were cut out from under me in practice and my left knee snapped like a stalk of celery.

I spent a week in the Cleveland Clinic having part of my patellar tendon grafted into where my anterior cruciate ligament used to be. Today, this surgery is a quick fix. But in 1990, it involved a lengthy hospital stay and some huge screws into your tibia and femur. Even before I talked to the recruiter, I knew that people couldn't join the military with metal hardware in their legs. So, my sophomore year, after I had rehabbed, I had a second surgery to have the screws removed...just so I could enlist without any problems.

As soon as he heard the letters "ACL," the doctor closed my file abruptly...never a good sign. "But haven't had a problem since then" I added, trying to undo what had obviously become a big problem.

"We don't take people with ligament surgeries." It was not an explanation, or an apology, or anything said with any degree of sympathy. It was just a fact. It was the way it was and now I was holding up the line. "Next!"

I walked out of line to my recruiter, who was standing in his Marine dress uniform with another sergeant. I tried to muster as much courage and self-respect as I could, but I was devastated. It was all I could do to keep from crying. I handed my folder to him without saying a word.

"What the fuck?" he barked; and then started to instruct me like he would instruct a kindergartner on the concept of standing in line. Just as he was screaming at me to fall back in, I told him that I had been rejected by the doctor.

"What the fuck for?"

"I had knee surgery."

"Why the hell did you tell them that? You should have just lied!"

First, I didn't know my surgery was going to be a critical flaw, and second, the doctor surely would have seen the scar. I explained this to the recruiter in a less than respectful tone, angry at the fact that the recruiter was blaming me for my rejection.

Back in the early 90s, peace was breaking out all over the world. The Gulf War was over, and the Cold War was ending. The military was downsizing and thus becoming much more selective with who they would and would not accept. Apparently, they didn't need a skinny kid from Ohio with one bad knee.

That appeared to be the end of my Marine Corps aspirations. I struggled after the disaster at MEPS. My whole life's plan had vanished in an instant. I was lost. I felt ashamed. I felt as if my identity and my dreams had been taken from me. When you are 18, the little twists that life throws at you don't seem so little. A very unfortunate event seems like the end of the world, instead of just one small detour in life. That's how MEPS was for me for a while, the end of the world. And I didn't have a plan B.

As time passed, I started to move on. After every disappointment, a bit of time seems to give you some perspective and serves to dull the pain a bit. my life went in a new direction, and I came to accept the fact that I was not meant to be a military man. The events at that time helped me accept that reality more than a bit. Since there was no war, and apparently no prospect of war on the horizon, there seemed to be less of a need to serve in the military, and less reason for me to be ashamed that I was not going into the military. I came to an agreement with myself that it was okay to not have served if we were not at war.

The second part of that agreement, however, was that if ever our nation found itself at war, I would be a part of it, whatever the cost.

Life happened after that. I went to college, became a police officer, and married a very short, very Italian, very wonderful woman. In 2001, my life was amazing. My career was engaging and challenging. My family was now nearly perfect with the pending birth of our first child.

September 11th changed the lives of every American, for

generations to come, if even only a bit. For me, that change was more significant.

On September 10th, I think that I, like many Americans, had a type of complacent patriotism. I flew the flag, voted, and did all the easy things that every American was expected to do to support our country. On September 10th, I thought that this was patriotism, but on September 11th, it all seemed like window dressing. On September 11th, I saw real patriotism.

I saw it on the faces of the police officers and firefighters who ran into buildings moments before they collapsed, to save complete strangers. I saw it on the faces of the families of those same police officers and firefighters who didn't survive. I saw it on the faces of servicemen who worked heroically to pull their friends from the Pentagon as it erupted in flames; and I heard it in the voices of ordinary Americans who gave their lives to protect their countrymen, pausing only long enough to say, "Let's roll."

These Americans bore the face of patriotism that I saw on my father, my grandfathers, and without exception, every other veteran I had ever known. It was a patriotism born of war, and the sacrifices that come with it.

Patriotism is a virtue that I believe every American holds. We either receive this gift at birth, and it is fostered subtlety, throughout our lives; or we fall upon it in an instant when we arrive here from a foreign land and realize the wonder that our nation holds.

However, our individual patriotism is born, I think that there inevitably comes a time in each generation that we have the opportunity to express that patriotism with selfless sacrifice. For my great grandfather, it was in 1911; for my grandfathers, it was 1941; and for my father, it was 1970. For me, it looked like 2001 was going to be my year.

Immediately after we were attacked, the shame that I felt for not having enlisted was all consuming. I started thinking about the promise that I had made to myself and how I would deal with it. We were certainly at war. I had no idea how long it was going to last but I wasn't about to renege on what I had promised myself on that day in MEPS.

I had never told my wife about my failed enlistment, about this now very haunting promise, or about my views on the military. It hardly seemed pertinent to our relationship…until shortly after nine in the morning on the 11th of September 2001. But all of a sudden, it became pretty darn pertinent.

It was easy as an eighteen-year-old kid to make some valiant promise that I would jump up and defend my country at the first shot fired in anger...consequences be damned. It was something entirely different to do that when I now had a wife, and a kid, and a job that was literally saving lives and fighting evil.

In the weeks following the attack, I was consumed with self-examination. What were my obligations to my country that was obviously at war? Were those obligations any more important than the obligations I had to my community as a police officer or the obligations I had to my family? There are thousands of men that can serve in uniform, but who will raise my daughter if I am killed?

I can't under state the emotional struggle that I endured when I was trying to decide whether to enlist or not. Joining the service would mean extended time away from my wife, my new daughter, my job. Everyone and everything that was a part of my life, that I had poured so much effort into, would suffer without me. And that's provided I could manage to go to war and come back in one piece. I had a duty to my family, to my career, and to the people in the city that I protected as a police officer. These were not responsibilities I could just walk away from. These were not things that would go away if I were to leave. This was my life. I was needed at home.

It's amazing how many valid excuses good men can come up with in times of struggle to justify to themselves that they shouldn't be the one to sacrifice, that they should leave that job to someone else.

But the fact was that Americans were fighting and dying quite literally to keep me and my family safe from harm. Didn't I shoulder the same obligation as an American to do the same for my countrymen? What makes their lives less important than mine?

So, there it is. A rather comprehensive list of very valid reasons to stay home, pitted against only one reason to the contrary. The choice to play it safe and stay at home seemed obvious, but it was not. The very essence of sacrifice is giving up something of value for a greater good. If enlisting wasn't so detrimental to about every aspect of my life, then it really wouldn't be a sacrifice...it would just be a job change.

The questions kept pounding away in my head. Just about every member of my family before me had made similar sacrifices in war for their country. What was so special about me that I got a pass this time around? Why should service to a country skip a generation? What would I want my children to do if they were faced with the same decision? What did it say about me if I took the easy way out?

And on the flip side, I wondered what would happen if I did enlist. Would my wife leave me if I went? Would I get fired from my job? Who would mow the grass when I was away? Who would walk my daughter down the aisle on her wedding day if I was killed?

I chose to enlist.

If I had not, the emotional conflict within me would have just continued, and I would never have found peace within myself. There was no way I was going to let good men go and fight a war for me while I sat at home. Even though enlisting would affect everyone in my life, the decision was one I had to make for myself. Even though it would be difficult, and would probably cost me a lot, it was the right thing to do. And I think that we should never pass up an opportunity to do the right thing.

One thing that I have to say that I am not proud of is the way I handled the matter with my wife. The journey to my decision was made much more difficult by me taking it alone. Only once did I even hint to Louise that I was considering enlisting. We were lying in bed, and I told her that considering the attacks of 9/11, I thought it might be my duty to enlist. She simply told me it was out of the question, that she would never be married to a military man, and that was the end of the discussion.

All of my turmoil, all of my reasoning, all of my debating was done in private. She had no idea of the war that was raging inside of me, or of my plan to enlist, until I had come to a final decision. I never spoke of it again until I had signed the papers. I didn't even tell her; my recruiter left a message on our answering machine telling me that my package was approved. Unfortunately, she got home before me and checked the messages.

She was devastated.

Of all the classless things I have ever done in my life, this was number one…and there is not even a close second.

Although I have told her since, I want to take the opportunity to do it again. Louise, I am truly sorry for distancing you from one of the most important decisions of our lives. It was thoughtless, inconsiderate and hurtful, and I am an ass for having done it. Thank you for forgiving me, and for standing beside me in spite of it.

I decided to join the Navy Reserves. Enlisting was no small feat since I had been rejected once already. I decided against the Marines, mostly because the training time was far greater than the other services, and because I was ten years older, and a few pounds heavier, than when I first tried.

I had worked my way through college as a paramedic, and still worked part time for my local fire department. I knew that the Marine Corps did not have their own medical personnel and that they used Navy Corpsmen to take care of their medical needs. I figured that was something well within my skill set, so I enlisted as a corpsman.

In the Navy, corpsmen functioned in various medical capacities: lab technicians, operating room assistants, dental technicians, and sick call screeners. The job of corpsman in the Navy was a jack-of-all-trades medical assistant who, after boot camp, attended a broad, basic medical school called "Corps School." In Corps School, prospective corpsmen learned everything from fundamental anatomy and physiology, to minor surgical procedures, and everything in between. After completing Corps School, corpsmen would then ship off to a Navy "C" school or finishing school for the corpsman's specific duty assignment. If the corpsman was going to work in a pharmacy, they would attend pharmacy training; if they were going to be an operating room assistant, they would receive advanced surgical training. There were "C" schools for just about every medical specialty in the Navy, including an assignment with the Marines.

Most of the folks working in Navy medicine are stationed in a fleet hospital, or on a ship; or provide medical care at clinics at various Navy installations. All of these services function pretty much in line with the way that civilian medical services work. There is a small segment of Navy Medicine that operates in direct support of the Marine Corps, and this was called going to the "green side."

The incentives the recruiter offered me when I enlisted were actually pretty attractive. Because I had a college degree, I enlisted as a Hospital Corpsman 2nd Class (HM2), or the equivalent rank of sergeant in the Army or Marine Corps. This was a nice little perk, as it normally took about four years to achieve that rank. My Corps School was also waived because I was already a paramedic and it appeared that there was quite a shortage of paramedics willing to work with the Marines. Oddly enough, the recruiter never asked me if I had tried to enlist before, and if he wasn't going to ask then I certainly wasn't going to bring it up. Things seemed to be going nicely. That was until the medical exam.

Just like eleven years earlier, I had to pass a medical exam before I could enlist. The trouble was I still had the same disqualifying medical condition I had eleven years earlier, an ACL reconstruction. I thought that this was a ridiculous reason to disqualify someone. I was in fine shape, I ran three miles a day, and I was on the SWAT team. I had not had one problem with my knee in years. It was nonsensical that someone would arbitrarily reject me for something that was so obviously not a problem.

So, when it came time for my exam, I used a little trick that the Marine recruiter suggested eleven years ago in MEPS to deal with insane and insurmountable bureaucratic problems within the military. I lied.

"Any medical problems in your past that I should know about like surgeries or chronic illnesses?" the doctor quizzed.

"Nope."

"Okay then, you look fine to me."

And that was it. Less than a minute. I had prepared myself to carry this lie out to the end. I thought that there would be some type of brutal medical interrogation, that I would have to explain my scar. I even had a secondary lie all set up in case the doctor started asking about it. If that didn't work, I was prepared to lay out my whole life's story and beg the guy to understand that it was my *moral obligation* to serve my country. I was going to throw myself at his mercy and plead for him to make an exception, just one exception, for me.

Instead, all I had to say was "Nope." He couldn't have cared less if I had one leg, let alone if both of mine worked properly. He seemed entirely disinterested in the whole thing. It was all very anticlimactic.

While I was shocked at the ease of it all, there really was a more pragmatic reason for the brevity of the exam compared to my last one. We were gearing up for a war. Men and women were already fighting in Afghanistan, and we would soon be fighting in Iraq and deploying troops throughout Asia and Africa to fight terrorists. And, as I would learn later during my pre-deployment medical exam before shipping to Iraq, as long as I was healthy enough to survive the flight overseas and get shot at…the military wanted me.

I checked in to my Navy Reserve Unit in Youngstown, Ohio, which was co-located with an Air Force Reserve refueling squadron. The first several months of my enlistment were a military "familiarization course" designed to prepare me for attending basic training. I was issued uniforms, attended classes in Naval history, and military tradition. I ran, marched, and ran some more. The pre-boot camp training was intended to funnel recruits into basic training at a higher state of readiness, and in turn, reduce the overall training time.

I went to Navy boot camp at Great Lakes, Illinois, in January of 2003, just as the rest of the military was preparing to invade Iraq. I remember it seemed surreal. I was in Navy basic training about twelve miles from downtown Chicago, nowhere near the ocean, where I would learn all about boats, in the middle of a bitter Chicago winter. All of this was in preparation for me to go fight a war on land, in a desert, with the Marines, who didn't have boats.

My boot camp was only seventeen days. Yes, it took only

seventeen days. The Navy crafted this program, called the Non-Prior Service Accession Course (NPSAC) for people, like me, who were older and already had extensive experience in their civilian skill set. All of us were in our late twenties or early thirties, and even though we were all enlisted, most of us had college degrees. My basic training company consisted of nurses, police officers, paramedics, an architect, a DEA agent, and an attorney. It was obvious that in the run up to war, the Navy realized that it had holes in many specialized skills that it needed to fight, and that running new recruits through the standard training process wouldn't put boots on the billets for another year at the least.

The NPSAC program was not designed to field combatant troops, it was designed to fill technical support billets. I read a statistic once that said for every infantry soldier on the ground, there are between five and ten folks giving that trooper the support he needed to do his job. The people in the NPSAC program were going to be those five to ten folks.

Before boot camp, NPSAC recruits were required to have completed a series of blocks of instruction so that when they arrived at Great Lakes, the seventeen days would be one final pass or fail culminating exam. Before showing up for basic training, each of us would have already been issued our uniforms and be able to pass a uniform inspection, be within height and weight standards, have passed the Navy physical fitness test, able to march, and have passed courses in Naval history, tradition, and operations. We were tested on each of the blocks of instruction we had taken before arriving and failures were sent home. The remainder of the time was spent in various team building exercises devoted to leadership, followership, and operating within the Navy structure.

Some sharp people graduated in my class and went on to serve honorably when called upon. And I think the NPSAC program is fine in theory, but I left boot camp feeling anything but military, and I am still ashamed to this day to admit to people that my boot camp was only seventeen days.

After boot camp, I returned to my reserve unit at the Vienna Air Reserve Station in Youngstown, Ohio. Each month, my training continued, and although I was in the military, I felt that my existence in the Navy Reserves was completely apart from the wars that were evolving in Iraq and Afghanistan. I was assigned to a fleet hospital unit, which was formed to be a mobile surgical unit for the Marine Corps and was equipped to care for massive casualties that would occur if we were engaged in a protracted war with the Soviet Union. My first training deployment with the unit was to the Marine Corps base at Camp Pendleton, California, where we spent the bulk of the week learning to set up a complex maze of tents which would house multiple operating rooms to deal with the casualties inflicted by the Soviets. I think it's important right here to point out that the Soviet Union hadn't existed for about twelve years, but nevertheless, we learned how to put up tents.

The fleet hospital unit that I was attached to was specifically not the Marine Corps and was in no way shape or form scheduled to deploy anywhere, let alone a war zone. I was not learning about combat casualty care under fire, I was not learning how to evacuate wounded Marines, I was not even renewing my CPR card. I spent my weekends giving vaccinations to new Sailors. This is not what I had signed up for after 9/11.

For about a year, I trained with my unit, while trying relentlessly to get assigned to the Marines. I was told that in order to be assigned to the Marines, I had to first complete Field Medical Service School (FMSS), the Navy "C" school for combat corpsmen, which was an infantry school run by the Marines for their medics.

Fine. Sign me up. When do I go?

But, like everything in the military, it wasn't that easy. In order to go to FMSS, I had to first be assigned to a Marine Corps unit. So, I couldn't go to the Marines because I hadn't been to their school, and I couldn't go to their school because I was not with the Marines. Wow.

Thankfully, I stumbled on to the secret of getting what you want in the military early in my career. I just kept asking different people. I got a different answer every time. Once I stumbled upon the answer I liked, I ran with it. Sometimes I didn't even have to ask different people, I just asked the same person a few different times. I'm always amazed at some of the benefits of disorganization.

The final person I happened to ask was Admiral Duret Smith, the Commander of Naval Reserve Medicine. I was at a Naval medical conference in Cleveland where Admiral Smith was speaking. Surprisingly, the topic of his speech had to do with the mobilization of Marine Corps infantry units during the initial Iraq invasion, and how in many instances there were not enough combat qualified corpsmen to stand up active duty infantry battalions.

The Naval Enlisted Classification (NEC) or job code for a Combat Corpsman with the Marines was 8404. Admiral Smith quoted one of the commanding Marine generals, who was preparing to deploy his division as saying, "What the hell is an 8404 and why is he fucking up my war?" Apparently, the lack of combat corpsmen was causing a serious issue for the Marines.

For every infantry battalion of about a thousand Marines, there were to be sixty-five corpsmen. In peacetime, back in the States, when no one was getting injured, a battalion might get by with twenty corpsmen. At any given time, we would have three Marine Infantry Battalions forward deployed, sailing around the world in Amphibious Ready Groups, ready to deal with any crisis that came up in the globe. Those would be the only battalions who had their full allotment of corpsmen.

Why? Because prior to 9/11, corpsmen serving with the Marines was a completely voluntary assignment. And there were just not enough folks who wanted to get wet, cold, and muddy with the Marines to keep those slots filled. So, when we tried to mobilize nine additional battalions simultaneously for the invasion of Iraq, and fill each of them with sixty-five corpsmen, the wheels fell off.

After his speech, Admiral Smith approached a small group of corpsmen that I was in and asked something to the effect of, "What can the Navy do to help you have a more fulfilling career?" The admiral was an amazing person, and had been the former Battalion Surgeon, or Chief Medical Officer, for the 3rd Battalion, 25th Marines, early in his career. He was an orthopedic surgeon in

Cleveland, and when he asked what he could do for his people, he genuinely meant it.

I raised my hand and told the admiral that I would like to go FMSS to help fill the void of Marine corpsmen and explained that I kept being turned down for the school because I was not currently assigned to a Marine unit. One email from the admiral to a personnel chief at Navy training and two weeks later, I was being screamed at by a Marine gunnery sergeant in Camp Lejeune. It was wonderful.

A testament to what a fine leader Admiral Smith was; after I returned from Iraq, he met our (his former) unit as we were demobilizing at Camp Lejeune. He welcomed us home, he wrote letters of commendation, he saw injured Marines in sick call. This was not his job. He commanded thousands of sailors across the nation, yet he took his own time to fly to North Carolina and take care of the Marines of our battalion, one last time. This was his last act before he retired. I had the opportunity to meet with and thank Admiral Smith during the demobilization process. I was chosen as the battalion enlisted medical representative to meet with the admiral to review some after action lessons. Admiral Smith always wanted to know how things could be done better. I thanked the Admiral for interceding on my behalf years earlier and allowing me to go to FMSS. He of course didn't remember, but I felt better for having done so.

Admiral Smith had served as an enlisted sailor during the Vietnam War, and after completing medical school in Syracuse, New York, he resumed his medical career with the Navy. He served in almost every command capacity in Naval Reserve Medicine. After retiring from the Navy, Admiral Smith continued to provide treatment to wounded Marines from our battalion, years after we returned. Rear Admiral Duret Smith was one of the finest Naval officers I ever met.

CHAPTER 4
FIELD MED

I reported to FMSS, or "field med," which was held at Camp Lejeune in January of 2004, one year after boot camp. I could tell instantly that field med was going to be nothing like the seventeen days I spent at Great Lakes Naval Training Center. First, this was not a reserve program; this was an active duty school. I was not attending with thirty something civilian paramedics, but with nineteen-year-old corpsmen straight out of corps school. This was not an abbreviated school designed to push corpsmen out into the field as soon as possible, this was a long, intense, detail-oriented school designed to keep us alive, so we could keep Marines alive. What everyone in this class was training for was quite literally a matter of life and death. These boys weren't training to go back to a Navy hospital; there were no non-combat assignments after graduation. All of them had already received their orders for after they completed field med: 2nd Tank Battalion or 2nd Marine Division or 2nd Marine Air Wing. These were all units that were forward deployed to Iraq; units that were already engaged in active combat and were taking casualties. Every single person who finished this school was going to war.

As I stood in line to check in, the heavy North Carolina air was thick with tension...and a bit of fear. One instructor put it succinctly,

"You joined the military in a time of war, which is where you will be going directly. It's our job to teach you to survive." The point was reinforced by our placement directly under rows of pictures of corpsmen who had all received the Medal of Honor…posthumously.

Of the ninety or so corpsmen in our company of trainees, only a dozen or so were asked to attend. The overwhelming majority had asked to be assigned or were already assigned to other hospital-based programs. Most of them had joined the Navy to learn a marketable civilian skill such as being an x-ray technician or phlebotomist and had intended to practice that job while serving their country. But in a time of war, the Navy draws its combat corpsmen not only from volunteers, but from all branches of its enlisted medical corps (and remembering that they were now desperately short of combat corpsmen) these young men were mandated to attend. In the years that followed, the shortage of combat corpsmen would become such a serious issue that the Navy made it mandatory that all corpsmen attend FMSS, just in case they were needed. I can't imagine joining the Navy thinking that I would be running an MRI machine in San Diego, only to be told that I would now be sleeping in the dirt, crapping in a hole, and getting shot at for the next year with the Marine Corps. Some of those kids had no idea what they were getting into. To be fair, I had no idea what I was getting into.

To say I was intimidated is an understatement. I was a twenty-eight-year-old reservist who had been exposed to active duty Navy life for seventeen days, surrounded by nineteen-year-old active duty corpsmen that had been immersed in the culture for a year. I had no idea what I was doing, and everyone knew it. I was an old man, a part time sailor, and more than a few of my classmates wondered, quietly and out loud, if I even belonged there.

As uncomfortable as I was feeling, it was about to get a whole lot worse for one poor fella. Just when I thought that I was the most embarrassing member of our platoon; in walked a bald, thirty something Jewish guy wearing his dress blue uniform. That uniform was only to be worn on the most formal of occasions, and certainly not to a combat school. He was also a reservist, and like me, had been sent to this school without the proper uniforms. He was late, he had missed his bunk assignment, and he had interrupted the instructors when he entered. Everyone in the platoon stared at him immediately. I might still be an embarrassment, but at least now, I wasn't alone.

The instructors quickly lumped the two old misfits together. After a few short minutes, we were talking. I looked at his name tag and tried to pronounce his name. After I slaughtered it, he corrected me in an atrociously thick New York accent, "It's Jeff. Jeff Wiener (WEEner)…like the balls." That was the first time we spoke, and for the next several months, we didn't stop.

Jeff and I found out that our military lives had been remarkably similar. We both joined the reserves after 9/11. Jeff was an EMT in Long Island who lost some friends in the attack on the World Trade Center. We were both married, and we both had new babies at home. We both tried to join the Marines out of high school, and both were unsuccessful because of some difficulties. We were both unimpressed by our seventeen-day boot camp and we had both volunteered for field med.

I think what really made us close though was that we both knew why we were there. We weren't complaining that we had been assigned there against our will like many of our classmates; we went there with a sense of purpose. We knew that our country was at war and both of us felt it was our duty to serve in a time of war. More than that, we realized that the Marines were at war, and we wanted in on it.

The first morning of field med, or any military school for that matter, is the worst. It's called "wake up morning" and it's called that for good reason. We woke up at 04:00 and met our instructors, who consisted of two Navy corpsmen and one Marine staff sergeant. During training, we would need to simultaneously hone our medical skills and learn to take care of wounded Marines under fire, as well as learn how to operate as infantrymen alongside the Marines. Giving us Navy and Marine instructors was meant to teach us both parts of our job at once.

As luck would have it, the school was short on Marines, so our platoon got three corpsmen. Hospital Corpsman First Class (HM1)

Banks, HM1 Bucchannon, and HM2 Tutko. These animals were three of the best instructors on the face of the planet. All of them were combat veterans of either Iraq or Haiti. Whatever an embarrassment the Navy was during my seventeen-day boot camp, it made up for it by producing these men. "Your mission," said our senior instructor, HM1 Jay Banks, "is to keep fingers on triggers. When a Marine is hurt, you have to get him back into the fight. You have to get his finger back on that trigger. And when you're not fixing Marines, your finger damn well better be on that trigger, too." HM1 Banks spoke with a slow, firm, North Carolina drawl. He didn't yell, he didn't scream, he didn't belittle; he just spoke the truth. When he spoke, it was like it was straight out of the Bible. He was calm, firm and honest, and when he said something, he meant it.

Wake up morning started with HM2 Tutko, a two hundred and sixty-pound body builder, throwing a metal trash can down the open bay that held our bunks and dumping a still sleeping corpsman out of his bed. We had five minutes to square ourselves away and be in formation for marching to chow. The first day I was a bit flustered and forgot to lock my locker. When we came back into our barracks to change into our running shoes and PT gear for the day, I found everything I owned dumped on the floor in the middle of the bay. I was sick. Instead of having five minutes to change, I had one minute, because it took me four minutes to find all of my crap and shove it back in my locker. Equally awesome was that HM2 Tutko was giving me words of encouragement the whole time.

"You are a fucking NCO in the Navy who is about to go to war! I'm telling you right now you better square your fucking self away."

I shot back a quick "Aye, aye!" and fell into formation. At the end of the day, we returned to the barracks. When we were marching towards it, I could see all types of clothing and equipment scattered in the yard and in the trees in front, as well as some things on the roof. I knew instantly that some poor son of a bitch had forgotten to lock his locker and had not learned his lesson from what happened to me. But when I got closer, I recognized one of the shirts in the tree. It was mine. I threw up in my mouth. I forgot to lock my locker. Again.

HM2 Tutko was standing next to the mess to greet us and explain to our platoon the gear scattered about the grounds. "Cleveland!"

If you are in the military and one of your instructors knows your name in the first eight hours of the course…bad things are going to happen to you.

I snapped to attention as HM2 Tutko crushed me in front of the platoon. "Fucking reservists! This is what it's come to? A year into the war and *this* is who we are sending to fight. Raise the white flag boys, we are going to lose this thing."

That first day was the worst day. Thankfully, I woke up the next morning and things slowly got better for me.

Field med was a ten-hour-a-day school divided into classes where we learned everything from treating sucking chest wounds to fighting through an ambush. We were immersed in Marine Corps culture, learning their rank structure, traditions, unit composition, and even singing the Marine Corps Hymn. Our Navy corpsmen instructors systematically went about drumming the Navy out of us and beating the Marine in to us.

They were meticulous about every detail: our fitness, our dress, our marching. Everything had to be perfect. One day, we had finished a three-mile run in formation only to stand for about twenty minutes of HM2 Tutko yelling at us for "looking like shit, running like shit, and acting like shit."

I didn't think we had done that badly. I was sort of happy with our performance and thought that it was one of our best up to that point. As a matter of fact, a platoon of Marines ran past us in a formation run right after we had finished, and they looked like a bag of ass. I pointed this out in protest to one of the other corpsmen in our platoon, nineteen-year-old Hospitalman (HN) Wilson, "Look at those jokers, we looked ten times better than they did." Wilson looked at me coolly and summed up the whole essence of Field Medical School in three words, "That's the point."

That was the point. We were not Marines, we were Sailors, and when we graduated and got to our Marine Corps units, we would not be expected to act like Marines, we would be expected to act better than Marines; to run faster, to look sharper, to be smarter, tougher, and harder. A corpsman was supposed to project leadership and give a sense of security. To do that, we had to be better. That's where we benefited from having three Navy men as instructors. The other platoons had someone teaching them to be Marines. We had three Navy men teaching us to be better than Marines; they were teaching us to be corpsmen.

That little attitude adjustment was just what I needed. I realized that anyone could pass any military school, but that wasn't the point. The point was to be excellent. Jeff knew this already, but I was a little slow on the uptake.

From that day on, I loved field med. I walked straighter, talked louder, dressed sharper, and worked harder. That school was what I had envisioned as a boy of what the Marine Corps would be like.

Because I had come into the Navy as an E5, I was a higher rank than about 90% of my classmates. Our platoon had three enlisted men who had been in the Navy for several years and were a higher rank than I was. The senior ranking class member was selected as the platoon leader and was given the instructional plan of the day to carry out. The platoon leader was in charge of everything we did: marching the platoon to chow, making sure that all of the platoon members were meeting academic standards, supervising the cleaning of the barracks, and making sure we pulled and packed the correct gear for field maneuvers. The platoon leader had the task of not only making sure that he maintained the high standards of the school, but that the other thirty corpsmen in his platoon also performed exceptionally; even though many of those corpsmen didn't want to be there.

Unfortunately, our first platoon leader did not do a good job. We were a hot mess, and he was fired within three days. The second platoon leader lasted a couple days longer but he got tossed, too. I'm not sure about the third platoon leader, but I think he was fired before lunch on his first day. I was lucky number four.

HM2 Tutko was so fired up that a reservist was now his platoon leader that he couldn't even form words properly when he yelled at me. It was mostly just spit coming out of his mouth and into my ear.

I could only make out, "I'm telling you right now," then a string of syllables followed by the phrase, "get yourself unfucked!"

And unfuck myself I did. I had already learned that the way to excel in field med was to do exactly what was asked of you, perfectly and immediately. I was past having any personal problems in training; I was ranked second in the class by this point. But being the platoon leader had zero to do with my personal accomplishments. There were still many, many sailors in my platoon that didn't want to be there and were not learning what was being taught. It wasn't a question of if they would pass the class or not (word had it that everyone was passing regardless of performance just so they could fill billets), it was a matter of if they would stay alive or not after they graduated. Those kids just did not understand the gravity of what they would be doing a few months after finishing school. From the time I took over, I made it my personal mission to try to impress upon them the seriousness of the situation.

With Jeff as one of my squad leaders, everything came together. Our platoon worked as a team and most of the malcontents decided that alive was much better than dead, so they started putting out.

One of our culminating exercises for graduation was a simulated mass casualty event. One of the other platoons acted as wounded Marines and fitted with mock wounds and fake blood. The remaining platoon would evacuate the wounded from the field and bring them to a makeshift field hospital for our platoon to triage, treat, and evacuate. I felt that our platoon had a very solid team in place. Of all the corpsmen in the whole company, only two of us had ever treated live patients before; Jeff and me, and combined, we had probably treated over a thousand. I assigned Jeff to be the triage officer, screening the cases to see who could be saved, and which patients needed to be seen first. Then I assigned HM3 Toya, one of our class's top radio operators, as the evacuation officer, and tasked him with coordinating the helicopter assets we would need to fly out our patients. I would be the medical officer, directing three treatment teams of four corpsmen on what to do with each patient.

From the initial patient, things went flawlessly. Everyone worked smoothly as a team, and there were no issues when the training staff started throwing harder and harder curveballs at us to see what decisions we would make. Forty-five minutes into the evolution, I looked up to see the entire training staff watching us, from the commanding officer (CO) of the school on down. I saw HM1 Banks smile as the CO told him that we were the best platoon he had ever seen run triage.

In the final stages of the event, the Senior Non Commissioned Officer (SNCO) of the school, a Navy Master Chief who had been a corpsman in Vietnam in 1970, came up to me as I worked to start an IV on the last patient and started to quiz me. "What would you do if that Marine didn't have any arms to start an IV?" Immediately, I shifted gears and started preparing to place a simulated IV into the jugular vein in the neck of the wounded Marine. The Master Chief stopped me, "Okay, but what would you do if he also had shrapnel wounds in his neck?" Without pausing, I removed the wounded man's boot and started preparing an IV site in the man's left foot. The Master Chief stopped me again. "You mean to tell me that if this man is bleeding to death, you can get enough fluid in through that little vein in his foot to save him?" I took a breath, and answered calmly, "Master Chief, once as a paramedic, I had a guy with tourniquets on both arms that were cut to the bone. I put an IV just like this in his foot and put two bags of fluid in. He lived. I promise you; this will work." The Master Chief looked at me for a full ten seconds without saying anything, probably trying to gauge if I was full of shit or not. Then he turned to the CO, pointed at me and said, "He's good to go. "

End of exercise. We killed it.

Just before graduation from field med, HM1 Banks and HM2 Tutko pulled me into the instructor's shed and told me that they had recommended me for Honor Graduate of the class. I was proud. There were several outstanding corpsmen in our company, any one of whom would have made a great honor grad. But for me to even be considered after picking my laundry off of the roof on day one; that was something.

Then, I was humbled. HM2 Tutko spoke next, "But we didn't give it to you because you sucked at pullups, it was embarrassing." I understood. The Navy physical fitness test consisted of a run, pushups, and sit-ups; all of which I crushed. The Marine Corps fitness test substituted pullups for pushups, and I was terrible at pullups. It took me two tries to get the minimum three required to pass.

I asked who would be named honor grad instead. My guess was a kid from 2nd Platoon in the Naval Special Warfare Combat Craft Crewman (SWCC) program. They were the sailors that went through a special selection process to drive the small boats that delivered SEALs on special missions. The SWCC sailors we had with us were very squared away. The SWCC kid did about twenty-five pullups without breaking a sweat…he was a lock.

Banks announced the winner, "Montoya, from 3rd Platoon." HM3 Montoya was a female corpsman who was five foot nothing, a hundred and nothing, and tough as nails; she was very deserving. I told the two that I was surprised, because I thought for sure the SWCC kid would win. Tutko gave one parting shot, "Nah, you didn't get beat by that stud, you got beat by a girl."

CHAPTER 5
WELCOME TO INDIA

After field med, I returned home and was assigned to a Marine Corps Reserve Infantry Battalion: Company I, 3rd Battalion, 25th Marine Regiment in Buffalo, New York, or India 3/25 for short. The battalion was split into four companies, each in different cities. Headquarters and supply in Cleveland, Ohio; weapons in Akron, Ohio; and three rifle companies: Lima in Columbus, Ohio; India in Buffalo, New York; and Kilo in Moundsville, West Virginia.

When I checked in to India, there was only one other corpsman in the company, HM2 Justin Mangano. Justin had recently left active duty after having served with a Marine unit. He was soft spoken, professional, and as hard as a coffin nail. He had been shouldering the burden of providing medical support for India Company solo as their training tempo increased for eventual deployment.

When I arrived, I fell immediately into company formation with 2nd Platoon. The platoon sergeant, Staff Sergeant Pudorhofski, a massive Russian who had just returned from an amphibious deployment in South America training allied forces, took one look at me and said, "I'll be. A fucking corpsman that's in shape. You don't see that very often." I appreciated to my very soul that HM1 Banks had taught us to "look tough, act tough, and be tough," because first impressions are very important.

Although 3/25 had not received official deployment orders yet, it was widely known throughout the Marine Corps that every infantry battalion, active duty or reserve, would be deployed to Iraq. The entire Marine Corps, reserves included, were on a rotational schedule to Iraq: every seven months, nine fresh infantry battalions would rotate into the country. The whole Marine Corps only had just over thirty battalions, so we knew our time was coming. When I checked in, we were only one of two that had not yet gone.

Normally, reserve duty is one weekend a month and two weeks a year of maintenance training duty. With an impending deployment, training funding was poured into our battalion. One weekend a month became Thursday through Monday. Two weeks a year became two weeks, three or four times a year, with an infinite number of individual schools that we had the option to attend. Headquarters Marine Corps was putting every effort into making sure that as reservists preparing to deploy to war, we would not be shortchanged on our training.

Over the next year, we would deploy to ranges in Fort Drum, Quantico, and Camp Pendleton. We rented abandoned warehouses and conducted force on force assaults with marking cartridges for our weapons. We spent hours breaking down crew served weapons, working on communications gear, and conducting immediate action drills. We were even scheduled to do a joint training exercise with the Dutch Royal Marines in Curaçao, but a hurricane put a stop to that.

I fit in perfectly with the India Marines. I was ten years older than all of them and my experience as a police officer and a civilian paramedic gave me a bit of credibility in their eyes. On one training evolution, we were riding in a bus to a machine gun range in Winona, Canada. I recall that the Civil War movie, "Glory," was playing on the bus ride. There was a scene in the movie where the 54th Massachusetts's Regiment was charging into a Confederate position in a line and were being torn apart by musket and cannon fire. Union soldiers were hit in the most graphic fashion. After one soldier was struck in the chest, a Marine asked, loud enough for the whole bus to hear, "Hey Doc, you think you could fix that?"

I shot back calmly, with a little smile on my face. "Easy. Sucking chest wound with an occlusive dressing. Done it a dozen times. You'll be fine." Everyone half laughed, not knowing if I was telling the truth or not.

The next soldier was struck in the abdomen and was eviscerated, his intestines falling out. The next Marine yelled out, "How 'bout that, Doc? Can you fix it?" I realized now that while the first Marine might have just been busting my balls, the men on the bus genuinely wanted to know that if they were wounded, that I could take care of them. They were looking to me not only to gauge my competence, but for reassurance that they were going to make it home alive.

I answered, slower and more serious this time, "Yeah, bro. It looks worse than it is. You would be up and around in a week."

The game continued. The wounds kept getting worse and the questions kept coming. Gunshot to the face? I can fix it. What if I caught on fire? I can fix it. Whatever horrible injury they would yell out to me, I would calmly and firmly let them know that they were in good hands; that I could fix it.

Just then, a soldier on screen had his leg amputated by a cannonball. "Hey Doc, if I lose my leg, can they sew it back on?" I tried to be non-committal. I said that there was a possibility that it depended on the wound. The Marines weren't buying it, they could sense my hesitation and it only made them ask more and more questions. Finally, Staff Sergeant Pudorhofski yelled out, "Listen jerkoffs, if your leg comes off in Iraq it's staying off. Suck it the fuck up and get a robot leg."

The bus went silent as we all thought about that reality for a minute. If your leg came off in Iraq, it was for sure staying off. The very next scene in the movie was a soldier being decapitated. A lone voice in the bus called out sadly, "Doc, I bet you can't fix that."

"No man, I can't." In fact, I knew I could not fix everything. In spite of how capable and confident I was, in spite of what I might say, they knew better. And soon enough, I would too.

Everyone on the bus sat motionless, staring ahead, digesting that they were not immortal, and that there was a very good possibility that when we went to war, they could be killed, or horribly maimed. The silence was broken by another Marine making a statement about this hypothetical severed leg. "Doc, if I lose a leg, just let me die. I don't want to walk around like a fucking pirate with a peg leg."

Another Marine made a quick correction to that statement. "You're a fucking retard. You can run a marathon with a robot leg. If you lose your arms, that's some shit. Doc, if my arms get chopped off, just shoot me, okay?"

Not to be outdone, another Marine spoke up, "Y'all are idiots. I don't care if I got no arms or no legs, as long as my dick is good, I can still fuck. But Doc, if my junk gets shot off, you might as well kill me right there." Everyone started laughing and the mood returned to normal.

As everyone started talking again, the Marine that was sitting next to me grabbed my shoulder and whispered quietly to me, so no one else on the bus could hear, "Seriously, Doc. I can't live in a nursing home or a wheelchair. If I get hurt that bad…" He took a minute to catch his breath. He was almost crying as he spoke again. "Don't let that happen to me, please. Don't make me live like that."

I grabbed the hand that was on my shoulder and looked at him quietly, for a long time. "I got you brother. No worries. I got you."

I kept in contact with my friend Jeff Weiner from field med regularly. Jeff received his own assignment with the 4th Tank Battalion in Fort Knox, Kentucky. We spoke often, hundreds of miles apart, about our units, our lives, and our training. Each conversation started out the same; "Any word on when you're deploying?" Neither of us had heard for sure that we were going to Iraq, but we knew it was a certainty.

In September of 2004, I got the word. We would leave after Christmas. Each infantry battalion was allotted sixty-five corpsmen. We had about forty. My chief, Anthony Horton, a soft-spoken Georgia boy and veteran of the first Gulf War with 2nd Marine Air Naval Gunfire Liaison (ANGLCO), gave me the heads up; "If ya have any friends that are hard chargers and want to transfer into the unit, we could sure use 'em. So, get em here quick."

3/25 would deploy with sixty-five corpsmen, forty of which would be ours. The other twenty-five would be a mix of corpsmen from other reserve units and active duty augments from Navy hospitals. Chief Horton was all about finding the best corpsmen for the battalion and he tasked us with reaching out to our friends who might want to deploy with us. It was not as easy as it sounds.

But those hard men did exist. Men who joined for the sole purpose of going to war. And when we found them, they were worth their weight in gold.

I called Jeff about two minutes later. "Jeff, were going after Christmas, are you in?"

"Hell, yeah!" he screamed in his thick Long Island accent.

Two weeks later, Jeff and I were standing in the freezing rain in the parking lot of 3/25, hugging each other and preparing to go to war.

September to January comprised near nonstop pre-deployment training, called a "work up," which was meant to provide Marines with more theatre specific skills for use in combat.

The first training for the two of us, and several other corpsmen from the battalion, was a special operations medical class called OEMS. The class was held in an aircraft hangar at New River Air Station in North Carolina. In classic military fashion, we were told not to discuss what went on there, because there might be some people that disagreed with it.

The class, instructed by Army Ranger and Air Force Para Rescue paramedics, taught us everything from advanced surgical procedures to how to perform field blood transfusions. It was a completely hands-on class and we used each other to practice. We would intentionally dehydrate each other to simulate head casualties, we would intentionally over hydrate each other to simulate an electrolyte deficiency, we would tie tourniquets on each other's legs and check the blood flow to the foot to see which one was the most effective; we would insert IVs into each other's arms and necks and bones and femoral arteries. Corpsmen were bleeding, passing out, and throwing up...and we fixed them just as fast as they dropped.

We had been at war in Afghanistan for three years and Iraq for just over a year at that point. In peacetime, medical practitioners often debated the best trauma and battlefield wound care, never knowing if what they taught worked. In a very short time during a war, the military had learned some hard and valuable lessons about what worked, and what did not. Some sharp folks distilled those lessons into a new program that would eventually be called Tactical Combat Casualty Care, which would eventually be pushed out to the entire military. But when it was first being taught to us, it was brand new and not accepted by the rest of the medical field.

We were taught how to make more effective tourniquets than the ones we were issued, and how using them outside of the recommended guidelines that could save lives. We were taught how to use new blood clotting agents, how to use plastic wrap to seal burns, and even how to start IVs in people's bones when they had no usable veins. All of these things are widely practiced now, but back then, the motto of the course was, "The lunatic fringe of medicine."

The culmination of the week-long class was a field exercise in the hills of North Carolina, behind one of the active duty corpsmen's houses. It was a "live tissue lab."

In plain English, the instructors were going to take a truck load of pigs and blow them up; and it would be our job to keep them alive. Pigs were set on fire, shot with AK-47s, blown up with explosives, and stabbed. In fairness to all the animal rights people reading this, there was a veterinarian on hand to anesthetize the animals before they were wounded.

Of all the medical training I have ever had, the live tissue lab was by far the best. To be able to practice advanced surgical techniques on real living creatures with real wounds: the experience was invaluable. Hundreds of corpsmen would take this course over the duration of the war, and I truly believe that doing so saved the lives of at least as many Marines.

After the live tissue lab, Jeff and I went out to dinner. We were excited for the training, but the lessons we learned did not sit well with us. In March, live Marines, many of whom would be only teenagers, would replace the pigs. Jeff and I both knew it, and our conversation turned somber.

"Do you have your will done?" Jeff asked.

"Yeah, Louise and I sat down and figured it all out, for both of us, so things were set even after I got back. We figured out who would get the kids, the house, life insurance, the whole nine yards. You?"

"Just my military will. I need to sit down and do all that."

We talked a bit more about the things we were doing to prepare our families. Who was going to shovel the driveway after it snowed, who would fix stuff around the house, what we were doing for health insurance.

Then Jeff asked me a question, it was sort of innocuous at the time. But looking back, not so much now. "If your wife died, would you remarry?"

I took a long drink and thought about it. "Naw. I'd just be fine with me and the kids. It would be kind of hard, you know."

Jeff agreed. "Yeah, me too. I don't think I could marry someone else."

Jeff followed that up with "If you died, would you want Louise to get remarried?"

"Sure. I would be completely fine with that. I don't want her to sit around with a couple of kids the rest of her life crying cause I went and got myself killed. You?"

"Yeah. Maria's a good girl and a great mom. She's a good judge of character, she wouldn't pick an idiot. I would want her to be happy. I trust her." There was a finality in that when he said it.

CHAPTER 6
GOING TO IRAQ

I received my orders to report to 3/25 no later than January 3rd. The call finally came after three different false alarms. At the end of October, we were told we were going December 1st. Then right before Thanksgiving, we were told that we were not going at all. Then it was on again, then it was off. The last three months were beyond frustrating. Planning for someone to mow our lawn, then canceling it. Passing off my cases at work, then showing up the next day. I didn't know if it was intentional, but by the time the order came, I could not wait to get to Iraq.

The day that I left, my wife drove me to Buffalo. I didn't trust her to drive alone on the way home, so my mom came with us. During the two-hour trip from my home to Buffalo, Louise and I barely spoke a word. She cried and I tried not to throw up. My wife, who was six months pregnant, was having a nervous breakdown and I was beating myself up for having caused it. Things had not been great between us since she found out I was deploying. Heck, things had not been great between us since I joined the military, but the notice to deploy took our marital dissatisfaction to a whole new level. We fought often, about little things mostly, but the root of all of our arguments was the war.

Now, as we stood in the freezing rain outside of the Drill Center, with only minutes left together, I felt ashamed that I hadn't risen above all the petty bickering and used the few months we had in a more meaningful way. I felt ashamed I was leaving my wife, my daughter and my unborn child this way. I felt ashamed of the decision that I had made. I had been defending myself, and my decision, for two years. Now, I couldn't have felt more wrong for having made it.

I kissed my wife goodbye and hugged her for what seemed like an hour. I realized that for as long as I stood there and held her, Louise would stay. She literally would not leave my side. I realized right then and there that she was not leaving me, ever. That no matter how much she wanted to hit me with a car, she would take care of the family while I was gone, and of me when I returned. That's loyalty… and I was abusing the hell out of it.

I gave her one final kiss. Told her I loved her and that I would see her soon, turned and walked away. She stood there crying, her pregnant belly covered in snow. The hardest thing I have ever done in my life was to leave her and not look back. She was my girl, and always would be but for the next year, that life would have to be behind me.

From Buffalo, Jeff and I flew together to Camp Lejeune. We flew commercial air from Buffalo to Jacksonville, North Carolina…with all of our equipment that we were to go to war with, except our weapons. As we passed through TSA, security had a fit looking at the x-ray of our bags.

We had a debate with the screener about our scalpels, cans of sterno, IR strobes, knives, flares, and even a severed hawk's foot that my dad had given me for good luck. At the end of the day, the pile of stuff that we were not allowed to bring on a plane was so big that they finally gave up and just let us check our bags.

All the corpsmen in the battalion deployed a couple of weeks ahead of the Marines to Camp Lejeune to allow us to sort out some Navy paperwork before officially being placed under Marine Corps control. That would be the first time that I would meet most of the other corpsmen in the battalion that I would be working with on that deployment. I made some new friends, like Jose Santos, a Cuban nurse and reservist from Florida who had been deployed from another unit to fill out our quota of corpsmen. And I reconnected with some old friends from the headquarters company in Cleveland, Jimmy Aluni and Vic Lewis, both firefighter/paramedics from the Cleveland area.

Collectively, they were a great bunch of guys, real professionals. Most were civilian firefighters, paramedics, nurses, flight medics, SWAT team tactical medics, and physician's assistants. At this point in the war, in an active duty infantry battalion, you maybe had ten officers and corpsmen that had treated live trauma patients before deploying. Of those, maybe two or three had done it on a regular basis.

For the reservists that made up the medical staff of 3/25, trauma medicine was everyone's full time job. Everyone was familiar with treating gunshot wounds, amputations from car crashes, and packaging patients for evacuation by helicopter. The level of medical experience assembled in our battalion was, with no exaggeration, the best of any infantry unit in the United States military in 2005. No one else even came close.

The battalion had also received a group of about a dozen corpsmen from the Naval Medical Center at Great Lakes, Illinois. All those corpsmen had previously been involuntarily sent to field med and had been conducting their normal jobs in the hospital when they were notified that they would be deploying to Iraq with the Marines. None of them were happy.

From Camp Lejeune, we flew to Twentynine Palms, California where we joined the Marines of 3/25 for two months of live fire training, designed to take the four separate companies from our battalion and teach us to fight as one group. A Marine infantry battalion normally has about a thousand men, but prior to our deployment, we had somewhere around seven hundred. The fact that we were under strengthened was what caused us to be the last infantry battalion in the rotation deployed to Iraq. To fill our remaining slots, Marines from two other reserve battalions (4th Reconnaissance Battalion and 2nd Battalion, 14th Marines) were deployed to augment us. The recon Marines were not set to deploy as a unit and were only deploying as individuals to fill out other units. That made us the recipient of a couple of hundred highly trained, and highly motivated Marines from bases in Alaska and Texas. The 2/14 Marines were from an artillery battalion. Since we were fighting primarily in cities, the Marine Corps had a very limited artillery presence in Iraq. There was just too much risk in firing cannons into cities, we would almost certainly kill civilians. Since those Marines were not needed to deploy with their artillery battalion, they were sent to an infantry battalion that was deploying.

The training was the best the Marines had to offer, complete with entire towns constructed to hone our urban fighting skills, real Iraqi immigrants used to simulate the culture we would see overseas, and live fire ranges that allowed us to sharpen our weapon skills.

We shot thousands of rounds, threw dozens of hand grenades, called in air strikes, fired our mortars and assaulted bunkers under cover of machine gun fire time and time until we could do these things without thinking...in the dark…with no sleep.

When we left Buffalo, India Company left with three corpsmen: Jeff, Justin Mangano, and me. For some reason known only to the Navy, Justin was sent to another unit in an area near Fallujah, leaving us with only two medics for the company. That would be the last time I ever saw Justin. He served honorably in Iraq and returned shortly before I did and started training as a New York State Trooper. But before he finished the police academy, he took his own life. For some, they can leave the war, but the war never leaves them. Justin Mangano was a good man.

When we arrived at Twentynine Palms, we were assigned our additional corpsmen. HM1 Glenn Minney, who had prior active duty service on a Navy ship, and was a pharmacy assistant for the VA in Chillicothe, Ohio. He was assigned as the senior medical representative for India Company. Also assigned to us was HM2 Vince Goekler, HM3 Matt Coleson, and HM3 Travis Wilridge, all active duty corpsmen from Naval Hospital Great Lakes.

Between field exercises, there would be break out training sessions. Jeff and I would teach medical classes every chance we got. Our training was focused on two areas, one, training Marines in our unit as "Combat Life Savers," or Marines trained to help corpsmen in treating casualties: and two, conducting training within the medical staff of India on standard operating procedures for casualty treatment, reporting and evacuation.

There were also break out classes taught at the battalion level. As comfortable as I felt with my medical skills, I wanted to take the time in Twentynine Palms to learn all that I could about the other side of "fingers on triggers." Calls would go out to the companies to send all of their radio operators to an advanced communications class for two days. I talked to the First Sergeant and made a case that I should also attend, telling him that if I had to call in a medevac, I would need to know everything I could about how to operate a radio. When I went, I was the only corpsman there. I left that class knowing how to work the three types of radios we used, operate the air to ground frequencies, and activate the repeater system and change the cryptography.

When a call went out for Marines to learn to call in close air support, I made a case to the company First Sergeant that I, too, would need to learn to talk helicopters into landing zones. I left that class not only knowing how to talk to medical evacuation helicopters, but also how to call in fixed wing air strikes, rotary wing close air support, and how to mark targets with tracer rounds.

Every single class that came down, I found an excuse to go. Sniper familiarization. I might have to go along on a mission to support them. Advanced improvised explosive device (IED) course; I'm going to be treating IED blasts, I might as well know the ballistics behind the bombs. It was a running joke in our company that I might have been the only corpsman to weasel his way into every one of those combined arms classes.

In our own battalion's course work, one training session that Jeff and I taught was about the advanced tourniquet use we learned in OEMS. Each Marine carried an Individual First Aid Kit, or IFAK, that contained the medical supplies he would need to either treat himself or a fellow Marine in the case of an injury. In the IFAK was a piece of rubber tubing that was to be used as a tourniquet. We had heard horror stories about the rubber tube tourniquet in the opening days of the war. Marines and corpsmen were trying to use it to stop bleeding but found it completely ineffective. When it didn't simply break, they could not get it on tight enough to stop any bleeding. It was costing lives. We were told that the rubber tube was scheduled to be replaced with the more advanced Combat Application Tourniquets (CAT) that we tested at OEMS, but apparently, that had not happened yet.

Before we left OEMS, one of the Air Force PJs showed us a way to make an improvised CAT tourniquet out of a triangular bandage, a hand grenade pin, and a dowel rod. For two days, Jeff and I worked to reconstruct this better option. Our first stop was the hand grenade range where we sifted through the dirt for a few hundred grenade pins. Next stop, the Twentynine Palms Naval Hospital to steal a few cases of triangular bandages (Matt Coleson happened to know a kid who worked there) and then finally to the Home Depot, where we put two dozen eight-foot dowel rods on our Visa cards and had them cut to size. At the end of the day, every India Marine deployed with a tourniquet that could (and would) save their lives.

Although the training was great, the living conditions were terrible. We lived in dusty Quonset huts in the middle of the desert where the temperature ranged from two degrees at night to eighty degrees during the day. That, combined with living within six inches of another man for two months, put everyone's health in the toilet. A quarter of the Marines in India Company had pneumonia and another quarter had severe dysentery. The chow hall offered the same three meals for the entire stay. At one point, I truly believed that the conditions were intentionally this bad to make us that much more eager to go to Iraq. In the end, we found out that our living conditions were just a logistical oversight by our command staff. We could have been staying in empty, HEATED, barracks at the main base in town, if we had only asked.

The only saving grace at Twentynine Palms were the food trucks. No matter where in the desert we were, if we had hiked for two days or rode four hours in trucks, at lunch time, we would hear a vintage Mexican car horn and a food truck would come rambling through the desert. Those culinary heroes would drive for hours and dodge unexploded ordnance just to take our last six dollars and sell us a Lo-Mein bowl. For that, food truck man, I will be forever grateful.

As our training wound down, we all had some spare time on our hands. Time that was spent...well, gossiping and speculating. Mostly about what our mission would be when we got to Iraq and where we would be sent. We were all following the news as best we could, trying to keep up with where the Marines were fighting. Fallujah. Ramadi. The Syrian border. All certainly seemed possibilities.

As the days wore on, bits of information started to leak out. Some of the lance corporals in our company were remarkably good at discovering classified information, we dubbed them the "LCN" or Lance Corporal Network. Every night, several of the NCOs would sit down, listen to the rumors that the junior enlisted men had heard, combined them with the little bit of intelligence we received from our officers, and formed some type of estimate about our situation. As it turns out, we nailed it.

From what we could tell, our battalion would be sent to a city called Haditha, halfway between the Syrian border, which was the responsibility of another battalion in the regiment, and the city of Ramadi, that was occupied by a separate Marine regiment. Of key importance in our area of operations (AO) was a massive hydroelectric dam on the Euphrates River. Our headquarters company (H&S) and Lima Company would be based at the dam.

The next strategic mission for the battalion was to occupy the forward operating base (FOB) outside of the city of Hit, a major population center to the south, and where we could oversee the strategic supply routes through the area. Kilo Company was tasked with this mission.

Weapons company was divided into about ten Mobile Assault Platoons (MAP) teams. The MAP teams consisted of four Humvees with about twelve to fourteen Marines each. They were responsible for reinforcing Lima, Kilo, and H&S with heavy weapons, acting as a quick reaction force (QRF) in case of emergencies, and, in general, patrolling our entire AO in search of bad guys.

India Company was detached from the battalion and stationed as base security for Al-Asad Air Base, the largest, and only, Air Force base in the area, and home to about every important command center in our AO.

As vital as this mission planning was, it was all put aside for two solid weeks at the end of our training at Twentynine Palms to facilitate something much more important: getting the entire battalion liberty in Las Vegas before we left.

Yes, a thousand Marines who were about to go off to war and might not return, who had not even smelled a woman or drank a beer in four months; were going to Sin City. I could not see how in the world this would end well, but it was sure to be a good time.

Most of the Marines who were married, including Jeff, made arrangements for their wives to fly out and meet them. Those who were not married flew their girlfriends out and got married. Those without girlfriends…found some.

I decided against flying Louise out. She was ready to give birth any day. The flight would have sucked, and someone would have had to watch our daughter Isabella, who was three. It just wasn't going to work.

To be honest, though, I didn't want Louise to come. Our goodbye at Buffalo was a killer. It took me a long time to be able to separate myself from the emotional pain of leaving her and focus on the task of getting ready for war. I didn't want to put her through that again, but more selfishly, I didn't want to go through it again either. So, I took the easy way out. Not exactly a husband of the year decision.

Moral dilemmas aside, the trip was a good time. There were the expected shenanigans; Vail got kicked out of Margaritaville for swimming in the fifteen-foot-tall margarita glass, Flatau got married, Hamilton won almost $10,000 gambling and blew it all in one night on booze and women, Weaver stole another Marine's hooker, Havens got VD, and everyone got into drunken fights. It was almost like we were following a movie script called "Marines in Vegas Before a War."

Jeff and I didn't see each other until the bus ride back. "How did it go?" I asked. I was happy for the time he spent with his wife and proud of him for being strong enough to make the decision to do so in the first place.

"Really great." He looked, and sounded, completely content as he told me about how wonderful it was to reconnect with his Maria and how appreciative he was for the time he was able to spend with her.

"I just couldn't bear having to say goodbye to Louise again," I caught myself explaining why I hadn't taken advantage of the opportunity to see my wife.

Jeff smiled from ear-to-ear and said "It was hard…" He paused for a moment, still grinning, "It was really hard to say goodbye."

Jeff taught me a valuable lesson about relationships. I never, ever, sacrificed time with my wife, no matter how short, no matter how hard the goodbye. I just don't know when it may be my last.

Our first stop after leaving California was Kuwait, where we would change from a civilian chartered aircraft to a military C-130 for the final flight into Iraq. We stood in chalks of twenty Marines and received emergency instructions from the crew chief, a forty-year-old tech sergeant with the North Carolina Air National Guard. The prop wash from the plane, his incredible southern accent, and the fact that he had his mouth stuffed with Red Man made everything he said a bit hard to understand, but this is what I did hear: "Now, we don't expect to be takin no fire on the landing, but if we do get shot down n' crash yall git out the plane and git as far away as ya can case this thing blows up! Ya hear me?"

Fantastic.

Well, we didn't get shot down, but what we saw when we landed on the tarmac at Al-Asad was every bit as frightening. A chalk of twenty Marines from the 1st Battalion, 7th Marines, who had been fighting along the Syrian border, stood waiting to load for home as soon as we got off. I would have thought that they would have been smiling, yelling, and jumping for joy at the prospect of getting out of Iraq. It was exactly the opposite.

They didn't look happy at all. They looked hollow, like ghosts. Their eyes were empty, their bodies broken, and I could just tell from the way they stood that they were emotionally devastated. Most of them just stared at the ground. A few of them looked at us as we ran past. They looked scared, terrified actually. I could not for the life of me understand why this wasn't the happiest day of their lives. Seven months later, I would.

Over the first three days, we assembled in the India Company berthing area, which was a collection of cargo containers specifically modified to house four to six Marines at a time, complete with air conditioning, electricity, and a TV if we wanted one. Al-Asad was technically in Iraq, but until that moment, they had suffered only one KIA. That was pretty good for a base of about ten thousand Marines. Statistically, we had a better chance of dying in a traffic accident in Cleveland.

Jeff and I met with the senior India Company corpsman, HM1 Glenn Minney, to see what our operational plan was going to be. Glenn had a peculiar personality and a bit of a nervous streak, but he was very knowledgeable and a very personable guy. Since the entire company was just going to be sitting on guard towers all day, not engaged in any fighting, there really wouldn't be any casualties to take care of unless the base came under fire. Still, we wanted to know exactly what we were supposed to do. To my surprise, Glenn hadn't been given any instructions either. "Beats me, just hang out I guess."

Marvelous. Everything I had gone through to get here, and I'm just supposed to "hang out." Not that I wanted to see people get killed, but if I was going to be away from my family for a year, I wanted to feel like I accomplished something.

That night, Jeff and I were summoned to the India Company office to receive our assignment. The company gunny, Gunnery Sergeant Haney, informed us that all of the India Company corpsmen, except for two, would be re-assigned back to 3/25 at the

Haditha dam for distribution throughout the battalion, and would be leaving on the first convoy out in the morning.

"I need two docs to stay with us, so our medical readiness stays straight, and in case, by some miracle, we go on an actual mission. I'd like it to be you two, cause you're not bags of shit. But because you're not bags of shit, I also want to give you the choice to go out to Haditha and get into the war. If you do decide to go, you need to pick the two docs who are going to stay. I have only two requirements. First: the two docs can't be complete idiots. Second: one of them can't be HM1 Minney. Let me know in the morning, gents."

Glenn had a bit of a personality conflict with some of the staff at India Company and was in effect, being fired. But everything happens for a reason. Later in the war, Glenn would go on to overhaul the way the battalion accounted for medical supplies and was able to start a program that passed those surplus supplies on to Iraqi Hospitals. He did a lot of good for a lot of people until the day he lost his eyesight in a mortar attack. Glenn Minney is a good man.

Wow. How about that for a life decision? Stay at Al-Asad, sleep in air conditioning and come home in one piece GUARANTEED or ship out to Haditha and go fight in a real shooting war. Two months earlier, I would have made my mind up that minute, but after seeing the Marines from 1/7, I was scared as hell.

Jeff and I walked to the Post Exchange, or PX, discussing what we should do along the way. "I don't know," I told Jeff in a bit of a skittish tone, "maybe we should sit this one out. I mean, we both have families. I think hanging back might be the right call."

I had a bad feeling about going to the dam. Heck, I was downright terrified. Every decision I had made up until this point had put me closer and closer to losing my family. Maybe this time, I should make the safe choice.

Jeff nodded in agreement. "Wherever we go, we go together. Deal?"

"Deal." Through almost every phase of our military careers, Jeff and I had been together. There was no way we were going to split up after all of that.

We didn't find much to buy at the PX, but what we did find was amazing. Jeff's brother, Josh, was a sergeant with the 1/7 who was rotating out of the country from the Syrian border. As we walked up the steps to what was the closest thing to a Wal-Mart supercenter in Iraq, Jeff and Josh passed within inches of one another. They had not seen each other in over a year, well before Josh deployed to Iraq.

The two stopped, looked at each other almost as if they had planned on meeting here and hugged without saying a word. Jeff introduced me. "Taylor, this is my brother, Josh." I was amazed. I had no idea Jeff even had a brother in Iraq, and for the two of them, among the thousands of Marines arriving and leaving, to meet by chance…it was just wonderful. I left the two of them alone to reconnect, and walked back by myself, wrestling with yet another impossible decision where every choice was a loser.

I stayed awake all night thinking, praying, and watching a steady stream of Blackhawk helicopters fly wounded from the front lines into the base hospital. Miles away from here, Marines were fighting and dying, and I was laying in air conditioning, trying to decide whether to join them or not.

The next day, Jeff and I met and walked to breakfast at the dining hall. "How the hell did that happen?" I asked about the chance meeting with Josh.

"I knew he would be leaving about the same time we were getting here, I hoped to see him. Funny how things work out."

There was not a minute of the last day that I hadn't thought about what I was going to do, and now it was decision time. We hadn't spoken since seeing Josh, leaving each other alone to make up our own minds about whether we were leaving or staying. I waited for Jeff to bring it up and announce what he had decided. I waited, and waited, and waited. Nothing. Just silence as we walked through the sand.

I couldn't take it any longer. I blurted out my decision, "Screw it. Let's go." I couldn't live with myself if some of our friends got all shot up and I wasn't there to help. "I don't want to be the guy that goes to war and sits on the sidelines eating Burger King." "Thank God," Jeff said, exhaling with a sigh of relief. "I hoped you were gonna say that. I don't want to hang out here, this place is gay. Let's go pack our crap." We skipped breakfast and told Gunny Haney our decision.

"Who's stayin back with India?" Gunny Haney asked. Jeff and I had asked the two most senior corpsmen, Doc Coleson and Doc Geckler if they wanted to stay or go, not telling them our decision, or that they would be the only two. Both chose to stay. We didn't give the junior corpsmen a choice, they were coming with us, along with Glenn Minney. We departed shortly after that for our new situation at the dam.

The dam was actually a pretty nice set up. It had a kitchen, a gym, a phone center, internet access, a chapel, and the berthing areas were made from various portions of the inside of the dam. It was a little worrisome to be living in a functional hydroelectric dam, but we couldn't beat the fact that it was a full thirty degrees cooler inside than out.

Atop the dam was the Battalion Aid Station (BAS), staffed by our battalion surgeon, Captain Matt Gratton. In his civilian life, Captain Gratton was an emergency room physician in Kansas City. He was an excellent physician, with a calm demeanor that was balanced by a steely discipline when it came to doing things correctly. He had every trait I admired in a leader. Matt Gratton was one of the finest officers I ever knew.

The BAS was responsible for the initial care and treatment of casualties received from the front lines. Each unit would have a corpsman attached who treated the casualty at the scene and then evacuated them back to the BAS, where we provided a higher, more stable level of care before evacuating the patient to the field hospital at Al-Asad.

The first few days were filled with boredom as we staffed the BAS and waited for our first casualties. We didn't wait long. Our berthing area was two floors below the BAS and connected by telephone. The corpsmen that we replaced had one very explicit recommendation for us.

"Only use that phone in emergencies to notify people of incoming wounded."

We immediately ignored them and became accustomed to using the phone as a regular means of communicating without walking up and down two flights of steps.

This time when the phone rang, a hurried call came out immediately after we answered. "CASUALTIES!" Everybody in the berthing area scrambled up to the BAS and quickly divided into treatment teams, with me heading one team and HM2 Zimmerman, a six-foot-seven paramedic from Texas, heading the other. We were given a casualty briefing by Chief Schaser: two Iraqi civilians were inbound with gunshot wounds. One was a child who was shot in the head.

Zimmerman looked at me seriously, and in a no-nonsense tone informed me, "Taylor, I don't do kids." I shook my head in agreement. I hated kid calls too, but it was clear that Z was asking for a favor.

We worked quickly and efficiently for about fifteen minutes to repair two serious, but uncomplicated, gunshot wounds to the head of a four-year-old girl and her father. We stabilized them and readied them for immediate transport by helicopter to Al-Asad.

The girl that I treated was the same age as my daughter. She was adorable. Her hair was a thick dark brown that was almost black, and her eyes were just a shade lighter. She wore a pink dress that came down to her shins. The dress was speckled with blood around the collar.

Of the two, the little girl was the more seriously wounded. In spite of her wounds, she remained conscious as I treated her. She cried softly and kept reaching her hand off of the treatment table towards her father. I knew that little girl could not understand and did not care that I was working to save her life, she only cared about being with her dad, where she was safe and loved. Her father sat upright as his wounds were bandaged. He stared at his daughter, silent and unblinking as tears rolled down his cheek, realizing that there was nothing he could do for her but pray.

After the flurry of activity stopped, emotion started to get the better of me. I became furious. This is exactly why I wanted to be in

Iraq, to save little girls like this, and to kill the monsters that would do that to her.

"Who in the hell shot a little girl in the head in front of her father?" I yelled to a group of about ten corpsmen who were standing around the BAS, trying to recover from the explosion of adrenaline. "What kind of animals are they?" Everyone stood silently, also wanting to know the answer to that very question. I asked again, slower and more demanding this time, rage building inside of me, "Who did that to this little girl?"

The reply came calmly and quietly from Chief Schaser. "We did. United States Marines shot those people. Their car wouldn't stop for a patrol, and they didn't listen to the warning shots, so they shot them."

The jaws of a dozen corpsmen dropped to the floor. I shut my mouth and walked silently back to my berthing.

I had a lot to learn about war.

Back at the berthing area, everyone tried to digest what had just happened, but couldn't. As we sat trying to make sense of it all, the phone rang again. Everyone jumped out of their skin. Our minds were still reeling from the tragedy we had just experienced. As I grabbed my gear and started sprinting back up to the BAS, a million different scenarios of what types of casualties I would encounter ran through my mind. I said a quick prayer that I would not have to treat another child.

Doc Hill answered the phone as he was running out the door, "Hold up, everyone!" Hill shouted. We listened to Hill answer a few short questions from the BAS as he hung up the phone. "No wounded, they just wanted to pass some word: everyone's birthing areas need to be cleaned by noon tomorrow. Doc Hill, who was about twenty-eight, was a slightly round hillbilly from Oklahoma who was one of our active duty augments. Prior to joining the Navy, he was a sheriff's deputy. He felt, after 9/11, that even though he was serving his country as a police officer, he needed to do more. I respected that. Doc Hill served bravely for our entire tour, escaping death countless times. Unfortunately, he was killed in a car crash shortly after he returned home, leaving his wife pregnant with twins. Brian Hill was a good man.

My heart was racing; I could barely breathe. I looked at HN Palacio, a nineteen-year-old active duty Filipino corpsman and saw tears in his eyes. The Pavlovian reaction to hearing the phone ring after what we had just been through was every bit as frightening. We decided after that moment, no one was to use the phone unless it was to alert us to incoming wounded.

The flow of casualties remained steady. An IED blast, a sniper, civilian shot. The work at the BAS was important, and fulfilling, to a point. I was still not out with the line companies who were working outside the wire, pushing out on real combat missions, and that was where Jeff and I wanted to be. That was the mission we had trained for and wanted to join.

Jeff and I were lamenting our dissatisfaction with being at the BAS over a dinner of what appeared to be boiled woodchuck when I ended the conversation abruptly. "I have to use the phone. Louise is having her C-section in an hour, and I want to talk to her beforehand." I knew what time she was scheduled for surgery, and I was hoping I timed it right to talk to her before she went into surgery. Half of me, though, was dreading the conversation. I didn't know what to say. I wanted to be supportive to her, but I was worried that I would just be an emotional drain on her. It was hard enough giving birth alone; she didn't need to be reminded that I could also be killed at any minute.

Jeff ran upstairs with me into the phone center, he may have been more excited than I was. I asked politely to cut in front of the two-hour line and was graciously allowed. I called the hospital and was able to talk to her as she was being wheeled off to surgery. "I love you, kiddo. Talk to you soon."

"How long before you know?" Jeff asked.

"About an hour or so I guess." I didn't know how I was going to stand the wait. Not being there for the birth of my son was killing me and waiting to hear any news was twice as bad. Unfortunately, my wait was cut short. "CASUALTIES!"

With every call of inbound casualties, the phone center and

internet access shut down immediately for two days. No one wanted to take the chance of someone getting killed or wounded and one of their friends notifying their family by phone, before an appropriate notification could be made.

Several hours passed as we treated a round of relatively minor wounded in a makeshift hospital in the middle of a desert. At the same time, thousands of miles away in another hospital, my wife was the patient, and all I could do was wait and pray. As I washed my hands, a Marine from headquarters handed me a Red Cross message. "Congratulations, Doc! It's a boy."

Jeff gave me a huge bear hug and yelled in a voice that could only come from a man who lived in a house full of women "You have a SON!"

Captain Gratton asked if I had spoken with Louise yet. "Before she went in, but the phones are shut down now."

"Come with me." Captain Gratton walked me to his berthing and allowed me to use the satellite phone which was issued to the battalion in case of emergencies. All I had to do was stand on the roof of a bunker to get reception.

We talked for about fifteen minutes. Thankfully, we were able to let the joy of the moment overshadow my absence. Louise was so wonderful and so strong. "Well, what's the final decision, does he have a name?"

"Harrison Taylor Cleveland" I liked hearing my son's name for the first time. "Can't wait to meet him."

The birth of my son made all the griping Jeff and I were doing about being stuck in the BAS the day prior seem petty. Our time would come.

It came several days later when Captain Gratton announced that two MAP teams, K-Bar 10, and K-Bar 7 needed corpsmen. K-Bar 7's had been fired for poor performance and K-Bar 10 had just been formed out of some supply and artillery Marines to help supplement the ridiculous operations tempo required of the MAP teams in the battalion. Accordingly, Jeff would go to K-Bar 7, I would go to K-Bar 10. We were officially in the war.

MAP teams consisted of about twelve to sixteen Marines and a corpsman, divided among four Humvees. Each MAP team was led by a sergeant or staff sergeant, each vehicle was usually led by a corporal, and one of the vehicle commanders served as the assistant team leader. The MAP teams were given the call sign K-Bar, after the fighting knife that the Marines carried, and were numbered sequentially one through ten.

Each of the Humvees had a heavy weapon mounted in the turret. At least one of these heavy weapons was an M-2, .50 caliber machine gun that was capable of engaging targets a mile and a half away with its armor piercing rounds, or an MK-19, 40mm grenade launcher that could shoot the equivalent of sixty hand grenades per minute at a target the size of a man, a mile away. Some MAP teams had both an M-2 and MK19. The remaining Humvees had a combination of M-249 SAWs and M-240G medium machine guns, the difference being the size of the round fired. The M-249 fired a smaller round the size that our M-16s fired, and the M-249 fired a larger, 7.62x54 cartridge. Some MAP teams had Humvees equipped with a TOW missile, which was a wire guided weapon used very effectively to engage Iraqi tanks during the invasion in 2003. But in 2005, with all of the Iraqi tanks long since having been eliminated, they were not good for much else.

Our four vehicles were a mix-match of varying types of Humvees. For the first two years of the war, the IED threat consisted mainly of smaller explosives designed to kill a human. Towards the end of 2004, insurgents started building larger and larger IEDs and placing them on routes of vehicular travel, with the intention of destroying our mobile forces. In 2004, most Humvees and trucks didn't even have doors on them, and if they had roofs at all, they were made of canvas. With more and more Humvees falling victim to mines and IEDs, inventive Marines had started welding scrap pieces of metal over the door openings and on the floors of their vehicles to increase the protection. Floors of seven-ton trucks were lined with sandbags to absorb some of the explosive force of the blast from a mine strike.

Eventually, the Marine Corps invented an up-armor kit to retrofit all of the Humvees in Iraq with increased ballistic protection, including heavier blast proof doors to stop rocket propelled grenades (RPGs) fired at the side of the truck, more armor plating around the turret gunner to give better protection from snipers, and blast shields around the wheel wells to deflect the explosive force away from the passenger compartment when the wheel struck a mine.

Fitting all of the Humvees in Iraq with these kits was a slow process and most of the MAP teams had only one of the much heavier and much slower, up-armored Humvees. Each MAP team had either a high back Humvee, which looked just like a pickup truck with high metal sides on the back, or a seven-ton truck, which is a semi-truck normally used for hauling equipment. Since neither the high back or the seven-ton were up-armored, motor pool Marines welded armor plating over every square inch of the trucks until they looked like vehicles from "Mad Max Beyond Thunderdome."

The overall mission of the MAP teams, who fell under the control of the Weapons Company from Akron, Ohio, was to be the battalion's presence throughout our AO. MAP teams carried enough food, water, fuel and ammo for about twenty-four hours, but oftentimes stayed out for longer.

MAP teams would patrol main supply routes for IEDs and mines, transport detainees to Al-Asad Airbase for interrogation and processing, provide protection for logistical convoys supplying our firm bases or combat outposts, and conduct casualty evacuations when air evacuation was not available.

MAP teams would also respond to positions where insurgent mortars were fired from, a job called "POO patrol" (POO meaning point of origin) and engage insurgent mortar teams. They also acted as blocking or flanking forces during assaults on cities held by insurgents. Most of the time, though, MAP teams simply drove around the AO looking to pick a fight and fights were not hard to come by.

K-Bar 7, Jeff's MAP team, was the personal security detail for our commanding officer, Colonel Lionel Urqhart. They would take Col Urqhart, and the battalion Sergeant Major Wagoner, to the various firm bases and combat outposts in 3/25's AO, supervising combat operations. By the time Jeff had joined K-Bar 7, that team had already been engaged in combat multiple times. Jeff was replacing their original corpsman, a four-foot-eleven Vietnamese kid named Vang, who looked more like an action figure than an actual person. During several operations, Vang had not performed in a manner that was in keeping with the highest standards of the world's finest Navy.

Before his replacement, the situation between Vang and his Marines was so caustic, that one of the K-Bar 7 Marines, Corporal Schueler came up to Jeff and me at lunch and asked, "How many scorpions can I put in Vang's sleeping bag before it kills him? I don't want him to die, but I want him to go home so my Marines don't die." There was no playing around with Schueler. He was a deadly serious man who was made specifically to go to war. When he asked this question, he was literally plotting Vang's demise.

Jeff and I sat in silence, a little taken back by the fact that Schuler wanted to half-assassinate one of our fellow docs, and not quite knowing what to say. Jeff spoke first, answering slowly and carefully. "Well, that depends if you are using little scorpions or big ones. The little ones are deadly, stay away from those. But the big ones...I would say four. Four should mess him up pretty good."

That minute, Schuler sized up Jeff and the two formed a bond that only real warriors can. One of instant respect that happens when you talk to a man for thirty seconds and can tell exactly what he is made of and measure his character. Schueler never did plant the scorpions in Vang's sleeping bag, but he did demand that Vang be fired and that HM3 Jeff Weiner was to be his replacement.

K-Bar 10 was formed as an additional MAP team to augment the crushing operations tempo of the other nine teams. During our workup at Twentynine Palms, the battalion moved around our practice battle space in the back of seven-ton trucks. Those seven-ton trucks had to be driven by someone and for us, those Marines were the artillery augments from 2/14.

The Marines have a saying; every Marine is a rifleman. In the Marine Corps, it doesn't matter what role a Marine performs: cook, truck driver, computer specialist, or a helicopter pilot, EVERY

Marine goes through basic infantry training after boot camp and is expected to be a rifleman if needed.

The 2/14 Marines were from Mississippi, Alabama, and Louisiana. They were the kind of boys who didn't waive rebel flags for show, because their ancestors legitimately fought on the side of the Confederacy. They didn't need a symbol to tell us where they were from, they had their bloodline. The 2/14 Marines were chosen as transport drivers for the seven-ton trucks because they were familiar with operating the vehicles which were also used to tow cannons and artillery shells, so this was well within their skill set.

When we arrived in Iraq however, we didn't move infantrymen around the AO with trucks because of the IED threat. Instead, we used Amphibious Assault Vehicles (Amtracks) or "tracks" for short. The tracks were armored vehicles that could hold about sixteen Marines, who would enter and exit through a drop-down ramp on the back of the vehicle. The track had a turret on top, with a MK-19 grenade launcher and a M-2, .50 caliber machine gun mounted side-by-side. The tracks were designed to be launched from a ship at sea and carry Marines to shore for amphibious beach invasions. Afterwards, they were supposed to accompany those Marines as light armored support, about fifty miles or so inland. The tops of the track opened up, so Marines could stand up and give the vehicle 360-degree protection from anti-armor rounds, which would slice right through the aluminum armor.

Unfortunately, Amtracks were never designed to function the way the Marine Corps had decided to use them in Iraq. The Army had Bradley fighting vehicles that had been serving this purpose magnificently for two years. The Marines had metal boats with bulldozer treads. The tracks were highly susceptible to mines, IEDs and RPGs, and they were never meant for use in combat for years on end. Our tracks had been in Iraq since the invasion and had so many holes in them that they would sink faster than an anchor when we tried to float them.

To facilitate the transport of our infantry companies, we were given a section of Amtracks from the 4th Amphibious Assault Battalion from Jacksonville, Florida. 4th tracks, call sign "Gator," was a reserve component with Marines from Florida to Louisiana. Even though I was not a fan of their machines, the Marines from the Gators were the finest around. They would risk their lives time and time again to save the lives of the infantrymen; many of them giving the ultimate sacrifice.

Since their previous job had been taken by the Amtracks, many of the 2/14 Marines found themselves out of a job. But since they had trained extensively in vehicle operations during our workup, they were able to come up to speed quickly and form K-Bar 10, to which I was assigned.

CHAPTER 7
MOTHERS DAY

I had been with K-Bar 10 for about a month and a half, and we had settled into our battle rhythm. We were generally tasked with daytime patrols and would be out for about fourteen hours at a time, before we returned to refuel, rearm, and refit. During that time, we had been lucky with our combat operations. We had hit several IEDs and had a few light ambushes, but none of them had caused any real damage to our equipment and we hadn't suffered any injuries except for some light fragmentation wounds to the face of our .50 cal gunner, Lance Corporal Sponagule.

After the first month of operations, we had to adjust the organization of our team. Initially, our team leader was Sergeant Joshua McCall, an active duty Marine and veteran of the 2003 invasion. Sergeant McCall had earned a bronze star as a TOW gunner when his anti-tank team destroyed a platoon of Iraqi armor during the push to Baghdad. He had experienced a ton of combat during his first tour and performed admirably. When we first arrived in Iraq, before we had seen any combat, Sergeant McCall would tell stories about the invasion to Marines that asked. When he did, he would always start the scenario off with the phrase, "During the war…"

A typical exchange often began, "During the war, we were in our

MOPP (chemical) gear for three weeks straight," or "Back in the war, you were digging fighting holes every night." The phrase was meant to elevate his combat experience in 2003 and trivialize our current mission as just, "stability and security operations," not combat. Not real combat anyway.

The initial invasion only lasted twenty-six days, though, and Sergeant McCall was not in combat that entire time. Our current mission schedule was grueling, and we would take contact about half the time. Let me be clear, there were MAP teams that had it WAAAAAYYYY worse than we did, but we were putting in work, no doubt.

It was a grind. There were no days off and each week, the fighting picked up a little more. More Marines were killed and injured. We were barely into the deployment, and we could see the writing on the wall that we were going to be lucky to come out of this thing with our asses intact. That proved too much for Sergeant McCall. Every man, even brave men, can only take so much; and this was proving more than he could bear. Sergeant McCall was relieved and the assistant team leader (ATL), Sergeant VanVoorhis (Sergeant V for short) was placed in charge. This made the vehicle commander of my Humvee, Corporal "Smitty" Smith, a twenty-three-year-old Forest Ranger and tree farmer from Mississippi, the new ATL.

We also needed a new vehicle commander for Sergeant McCall's old crew. Since there were no more NCOs in our team, Smitty, Sergeant V, and our section's officer, Captain Moeller, met to choose which one of the junior Marines would take over as the new vehicle commander, or fire team leader.

After the meeting, Captain Moller came to me and asked, "Doc, can you take over Victor 3? (Sergeant. McCall's old vehicle.)"

I was shocked. A Navy corpsman being given operational control over a fire team of infantry Marines does not happen, especially in a war.

I wanted to be crystal clear what Captain Moller was asking of me. "Sir, are you sure you're okay with putting a Navy guy in command of Marines?"

He shot back a little sternly at my question, "Doc, I don't care who the fuck is in charge of my guys as long as he knows what is doing. I just want someone who can accomplish the mission and keep my Marines safe. I want the best man for the job. Period."

"How do the guys feel about this?" I asked.

"They recommended you," he reassured me. I was honored to have the job, I just hoped I wouldn't let anyone down.

It was Mother's Day. I had just finished talking to my wife, trying to wish her the best that I could over a barely working phone with a five second delay four-thousand-miles away. Talking to her on important dates was always hard. There was the loss and anger of not being there for her when she really needed it. Then there was the pure frustration of not being able to describe to her the emotions that I had boiling over inside of me. Without saying anything I could tell that she was all at once angry at my absence, confused over my decision to be here, and utterly terrified of what might happen.

I wanted to tell her that I understood those things and felt the same way. I wanted to tell her that more than anything I wished I could be there with her. I wanted to tell her that every minute of every day my feelings swung pendulously between pride about service to my country and shame for abandoning my family. I wanted to tell her all of those things, but I just didn't have the words. All I was able to muster was a curt, "Happy Mother's Day. I love you." We spoke for a while but not about anything as important as I had hoped. Before I left, I met with the florist at home that I usually went to and I gave her my Visa card and pre-ordered flowers for all of the important holidays that I would miss. At the time, I thought that that would be nice in case I was in some remote part of Iraq where I wouldn't be able to place the order. After a couple of months in country though, I started to think about what it would be like if I didn't come home. Would they still send the flowers even though I was dead? I hoped so, Louise would have liked that.

As I walked back to my bunker from the makeshift phone center, I saw Jeff in the stairway. He had just come back from a mission with his MAP team and was smiling ear-to-ear, which was in stark contrast to the last time we had spoken.

Two days earlier, we had eaten dinner together on the dam. Jeff was, like the rest of us, struggling with being in Iraq. Jeff's struggle was not that he didn't want to be in Iraq, but that he felt that he wasn't contributing. We had been here a couple of months and there were plenty of Marines in the battalion that had already experienced some hellacious fights. Since Jeff had been transferred onto his MAP team, they had yet to see any significant action. He wanted to get into the fight; he wanted to help. Marines were fighting and dying in this battalion, and to this point, he felt he hadn't done enough.

I told Jeff about a book that I read when we were in Twentynine Palms. It was written by a Marine Gunnery Sergeant who served in Vietnam. This gunny, then a sergeant, served towards the end of the war, and the action that his battalion had seen was quite limited. So limited in fact that he began to wonder if he would be the only Marine in the entire country not to see combat. While on patrol one day, his squad was ambushed. In the hellish minutes that followed, the sergeant was able to rally his Marines, drive through the ambush and destroy a superior force, without losing a man. When asked about his actions, he replied "I trained every day of my life for the past four years for this. Four years of training came down to three minutes of fighting. That was the only action I ever saw, the only fight I was ever in. But I won. That's the whole reason I went to Vietnam, to help my Marines fight through that one ambush."

I didn't know what the defining moment would be for Jeff, but I assured him that it would come, and that when it did, he would undoubtedly rise to the occasion. Our talk seemed to put him in a better mood, which was odd, because up to this point, the roles had been reversed, with him dispensing the advice, and me being comforted by it. We left after throwing away the remnants of dinner and went our separate ways.

When I next saw him, he was calm and seemed completely content, just like he looked after visiting with Maria in Las Vegas. He told me that he had just read a book in the rather well stocked closet that doubled as a library at the dam. The book was a biography of every victim of the 9/11 attack. Jeff said that he looked at the faces of every one of the fallen for hours and came to the realization that their death was his purpose for being here. He told me, that after reading that book, "It all makes sense now. Things will be okay." I left him feeling proud of myself for giving some advice that had obviously helped. I had no way of knowing how foolish and self-absorbed I was to think my story was what gave him comfort.

I left my three-hour phone event feeling whipped. To an ordinary person, making a phone call would not seem like work, but when factoring in the three-quarter of a mile walk from my bunker where my MAP team was housed in the dam, then the walk up ten flights of stairs, then waiting in line for two hours only to have my heart ripped out because I could only talk to my wife and not hug her, then add to that another pleasant walk back down ten flights of steps and another three-quarters of a mile sand hike, smack through the very spot where the insurgents had managed to land dozens of mortar rounds: it was a bit draining.

So, I went off to do what all of the other Marines in the entire country did when they were in an emotional funk. Lift weights. As a matter of fact, lifting seemed to be everyone's remedy for almost any problem, because any Marine who had any kind of problem tried to lift it away. Sometimes it worked a little, sometimes not. But in any event, I was getting into shape, so I guess there was an upside.

I went to the small weight room that was created in the motor pool near our bunker. The weights were stolen from the battalion weight room at the dam. Someone probably got tired of walking to the gym, so they decided to build one closer to home, piece by piece, with stolen weights. There was a lot of irony in that, and it certainly spoke to the mentality in the Marine Corps. Someone carried an entire weight room, piece by piece, nearly a mile at a time, ten stories back down the dam, so all in order to get in shape but not have to walk as far. Amazing.

I was in the middle of a set when Captain Moeller walked through and into the com room. He had a purposeful walk, which immediately broadcast to everyone that something was wrong. There was a whole country that we had to walk through, and no one walked like that unless there was trouble. Before I could ask the details, Captain Moeller said, in a deliberately calm tone, "There's a MAP team in a fire fight, get everyone spun up."

That one sentence was powerful enough to set in motion a beehive of activity. I ran to the bunker and made a general announcement to my MAP team to stand, that there was another MAP team in contact. I spoke with Smitty, our team leader, and told him what I knew.

We had just pulled a mission earlier that day escorting a convoy to Al-Asad Air Base and back. Although relatively routine, it still left me exhausted when we returned. So exhausted that there were the usual moans and gripes about cleaning the weapons, and refueling the trucks, and doing all sorts of mundane tasks that had to be done to get our gear back in working order for the next mission. We were not scheduled to go out again until the next afternoon at the latest, and a brief thought was given to letting the maintenance go until the next day.

Captain Gratton once told me that part of being a professional was doing the right thing even when we didn't want to do it. It was the right thing to make my guys prep and reset their gear before they were off; it was the right thing for me to do that to my own equipment; and it was the right thing for Smitty to make sure that the whole team had done so. That discipline paid dividends once we were called on in an emergency like that moment to go save some lives. That's the professionalism that all of the Marines in our battalion demonstrated every day. It was just the way we worked.

The other Marines on our team were busy dressing and checking out ammo for the crew served weapons, while Smitty, myself, and the other vehicle commanders tried to get what information we could from Captain Moeller.

Earlier that night, we had received some mortar rounds, which was nothing new. As a matter of fact, it seemed to be a nightly occurrence. Our operations guys were able to get the location of the POO, where the mortars were being fired from. Having discovered that location, a MAP team was sent to the east side of the river to deal with the problem.

Our AO encompassed terrain on both the east and west sides of the Euphrates River. The west side held all the population centers and was generally where we operated. The east side had nothing but landmines and we usually stayed away from there. The dam was the only crossing point for things larger than a camel for fifty miles up or down river, so if we crossed to the east side, we were on our own.

Once that MAP team arrived in the area where the mortars were fired from, they started taking fire from across the river to the west, from the town of Haditha. Driving from the east side of the river back to the west where the fire was coming from would have involved driving back to and over the dam, which would have taken hours. So, for support, a unit of small boats attached to our battalion to patrol the river was dispatched.

Two Marine Corps small boat teams supported our battalion, one downriver from the dam and one in the reservoir, upriver from the dam. The boats routinely patrolled the river looking for bad guys, but we also used them to insert teams of Marines for patrols, or snipers for over watch positions. The boats also had three mounted machine guns, one of which was a 7.62 mm "mini-gun," that sounded like a chainsaw and fired a crippling three-thousand-rounds a minute through its five rotating barrels. On that night, the boats were called upon to use those guns to support the MAP team that was taking fire.

Once the boat teams arrived, they also came under heavy fire from the same area on the west side of the river, specifically, the city of Haditha. It was clear by that point that there was a significant force in Hadith, and that this engagement was going to be for real. To surround the insurgents in Haditha, and provide some relief to the Marines under fire, a second MAP team, K-Bar 7, (Jeff's Team), had been dispatched to the west side of the river to flank the opposing force. My Team, K-Bar 10, was being assembled as support and would report to the Combat Operations Center (COC) at the top of the dam to escort an additional team of Marines into the city.

The base at the dam had two parts, the ground level and the top, and there was a two hundred-foot elevation difference between the two. To get from the bottom to the top, we had to leave the base via a series of switchbacks up the side of the hill, to where we could re-enter the base topside.

As our MAP team left the gate for the COC at the top of the dam, I could see a hellacious mushroom cloud to the south, in the area where Jeff's MAP had been sent. I didn't breathe for what seemed like hours. It looked just like the 1950s movies that showed the testing of an atomic bomb. God, I had never seen an explosion that big. A million thoughts went through my head: "Did we have air support up? No, I didn't hear the Forward Air Controller (FAC) on the net. Was that a main gun round from an Abrams? No, too big for that. Crap that was a car bomb. God, I hope it missed."

I started breathing again only when I heard the radio crack to life. One of the Marines from Jeff's MAP put to rest any question that myself or anyone else had about what that explosion was. "Bulldog, Bulldog, K-Bar-7. Ambush. River Road at the 8-0, returning fire." The stone calmness of that statement was the polar opposite of what was actually occurring just three miles south of us.

My MAP team rallied at the COC and received a quick situation report. Jeff's MAP was pinned down in an ambush on River Road at the 8-0 northing. Northings are little lines on topographical maps that we used on roads that ran north and south to track where we were at, since there were no street signs. There were at least four Marines dead and it appeared that the rest of the team was wounded. The tanks that had accompanied them were taking massive RPG fire and were in trouble. Jeff's team was hurt badly and in desperate need of help.

This left us in a precarious position. Lima Company, and a good portion of the battalion assets that were normally stationed at the dam, were conducting an operation near the Syrian border. Except for the units already committed to the fight, the only Marines left were my MAP team and Marines from the battalion who were not normally assigned to combat duties, such as communications, supply, planning, and operations.

A frantic call went out to organize what Marines were left into an effective fighting force. It was a motley crew to say the least. Cooks, clerks, supply Marines, truck drivers, and pretty much anyone with a rifle assembled at the COC and loaded onto Amtracks.

The situation at the COC was chaotic to say the least. Amtracks were lined up with Marines piling into them. Comm Marines were trying to make contact with Jeff's MAP team. NCOs were trying to get a count of our troop strength and just about everyone was trying to figure out what the hell was happening on River Road to K-Bar-7.

I could hear a staff NCO, who had clearly lost some of his equipment, yelling to no one in particular, "Where are my NVGs… who took my fucking night vision? Somebody better shit me my night vision right fucking now or were gonna play!" Broad, profane laced demands directed to everyone within earshot were a common occurrence in the Marine Corps, and although it might seem completely insane to normal people, it was surprisingly effective under normal circumstances. Right now, at that moment, however, was not the time to hold others accountable for this staff NCO's inability to keep track of his gear. It infuriated me. I felt like telling him to square himself away instead of yelling at Marines who were about to go into combat. I felt like telling him that there were Marines out there who were dying, and we needed to go help them. I felt like telling him to hurry his dumb ass up and get on the Amtracks so we could move out. I felt…like punching him.

Looking back, I realize that like many of us, he was terrified. Losing his NVGs, and his reaction was purely fear and stress. Who was I to fault someone for that? I was probably two hundred times more scared than he was.

I briefed Captain Gratton at the BAS and informed him that he would be receiving a mass casualty shortly. He nodded his head slowly without speaking. The expression on his face said that he was carefully and quietly planning every detail of what was to come. As I left, I could tell he wanted to say, "Be careful," but instead, he stood silent.

As I climbed into my Humvee and left, I could see the remaining corpsman and Captain Gratton moving quickly and deliberately to ready the equipment they would need for the worst night of casualties many of them had seen.

The radio traffic was a complete mess. Officers were trying to gain information from units in contact, units in contact were trying to describe what help they needed, our FAC was trying to figure out where to have airplanes drop bombs. There was more activity and information during the five-minute drive than any twenty people could digest.

The most amazing part of this complete mess though was the fact that everyone was almost preternaturally calm. No one yelled on the radio, no one yelled at each other, and all of the orders that were passed along were clear, concise, and specific. It was as if we were all just riding the subway to work. The only thing that would differentiate this from just another day at the office was the fact that I was sincerely about to throw up, I was so scared. And I am sure that I was not the only one.

As we were driving down River Road, I heard a Marine from Jeff's MAP, calling in the Casualty Report or caserep. A caserep was a report to the battalion of how many dead and wounded a team had received, and what type of medical evacuation they required. It was a report that was supposed to be given by the corpsman. The only reason that would not occur was if the corpsman was so overwhelmed by casualties that he could not break away to get to the radio or if the corpsman was a casualty himself. In either event, it was obvious that they needed additional medical help. I let Captain Moeller, who was in my vehicle, know that K-Bar 7 was in a spot and needed some medical help to treat their wounded. I asked his permission for me to stop and give them support.

Captain Moeller quickly denied my request, as there was still a significant firefight going on and I was needed with the rest of my team. Our column came to a stop on River Road, just north of where Jeff's team had been ambushed.

Everyone except the drivers and gunners of our vehicles dismounted. I say everyone, but it only amounted to about four or five Marines. As I stepped out of my Humvee, I was immediately paralyzed. I saw the seven-ton from Jeff's team stopped along the road, there was not one square inch of it that had not been shot, and it looked like it was on fire.

The back of the seven-ton was full of bloody Marines. It looked as if a giant hand had reached down from the sky onto the beaches at Iwo Jima and heaped the dead and dying souls into the back of that truck. Marines were moaning, some laid silent, some screamed in pain. I could hear one Marine's plea for help, "I'm bleeding a lot…I'm bleeding a lot."

Before the thought of rushing to their aid even entered my mind, a tank sped past me and escorted that seven-ton as it staggered away towards the BAS. I remember trying to say a prayer as it left. Every time I evacuated someone, I always said one "Our Father" and one "Hail Mary" to myself as they departed. It was just a quick way of asking God to take care of them since I no longer could. I started this routine when I first became a paramedic. I had done it probably five hundred times. But this time, I couldn't remember the words. No matter how hard I tried, I just could not remember the words to the two Catholic prayers I had said daily since I was in kindergarten. I was speechless, the only prayer I could remember was the dinner blessing: "Bless us, O Lord, and these Thy gifts which we are about to receive, from Thy bounty, through Christ Our Lord, Amen." So, I said grace as my mortally wounded friends drove off. Hardly appropriate.

My prayer was abruptly interrupted by a burst of small arms fire from the river. Marines quickly dove for cover along the road, as officers yelled for us to "get some dispersion," so we would not all be killed by the same explosion. I was taking cover behind a pile of rocks that looked like, until very recently, had been a house. Rounds were snapping over our heads from somewhere. I looked up to try and see where the fire was coming from, but it was impossible. Someone was shooting at our team, and I had no idea where the shooters were. I think that confusion was the scariest thing in battle. Not knowing where the enemy was, who was shooting at me or where I needed to shoot was terrifying and it always happened in the first minute of every engagement.

I remembered watching Civil War movies as a kid and thinking how stupid it was for two armies to stand in straight lines across from each other and shoot. I think at least back then though; they knew who they were supposed to shoot. It probably made things a whole lot simpler.

A Marine grabbed me, passing an order that had been passed from up ahead in front of us, "Don't shoot to the left, there's friendlies over there." What did he say? With all of the shooting, I could barely make out his words and didn't understand if he said "left" or "west."

"Don't shoot to the left, or don't shoot to the west?" I asked him "...cause that's two VERY different things." The Marine looked stumped. He was passing an order that had been passed down the line and he had no idea what direction we should or should not be shooting. He shrugged his shoulders in response and took cover behind a tree stump.

As I looked around, I could see a line of Marines taking cover on the side of the road behind everything from trees to rocks to other Marines. It was a pitch-black night, and the only light was from a burning vehicle to our rear. I looked at it and thought that it must have been the remains of the car bomb, as it was completely obliterated and ablaze. As I was studying the burning hulk, someone yelled back to me, "Does that have a turret on it?"

What did he say? I had no idea what the phantom voice from up front was talking about.

I saw the Marine who was yelling start walking towards me from the line of Marines up front who were taking cover. As I was trying to process what he was saying, he repeated his question impatiently.

"Does that fucking car, that's all blown up and burning, have a fucking turret on it?"

I looked closely at the fire and yelled back "yeah" before I even processed what that meant.

The burning wreck did have a turret on it. It wasn't the car bomb; it was a Humvee. A US Marine Corps Humvee. A Humvee from Jeff's MAP team and anyone in it was certainly dead. After I confirmed the identity of the wreck, the Marine who was asking turned out to be from Weapons Company, First Sergeant Sauers. He shot out a quick order, "He's still in there, get some security and give me some fucking cover."

First Sergeant Sauers was a wiry but strong Marine with thick glasses and a black belt that could torture anyone who mistook his mild, proper manner as a sign of weakness. First Sergeant Sauers was the epitome of leadership. His presence and ability made people listen to him. When he spoke, never once would anyone, even silently to themselves, question what he had to say…because he was always right. Even if he had been Private Sauers, people would still have listened to him because he was always giving the right orders. Leadership is action not position and people will always follow someone who does the right thing. First Sergeant Sauers was that kind of man and leader.

It was comforting to hear someone giving me directions. At the same time, I was horrified at the thought of a Marine still trapped in the burning wreckage. I had a purpose now, something to do, and that helped stem my fear. I grabbed a few other Marines and set up defensive positions as First Sergeant Sauers led a team of Marines to the crash site. They worked feverishly through the flames to pull out the remains of a dead Marine who had still been trapped inside. We were still taking small arms fire as they carried Lance Corporal Lance Graham to an Amtrak and placed him inside.

Gunny McIntyre, the Operations Chief, was at the head of the formation yelling for Marines to gather at his position. The rest of my team and I found Gunny Mac squarely behind a tank, which was placed in an alley between us and the small arms fire.

Wherever I went, I always felt better when there was a tank there. Just knowing that the insurgents didn't have tanks, or anti-tank weapons, and that no bullets were going to go through it, gave everyone around a feeling of invincibility. Unfortunately, we would soon learn that even our vaunted armor, the best the world has ever known, was not invincible.

Some months later, I was running on the top of the dam when I saw some Marines from Lima Company looking off into the desert with a spotting scope at a column of thick black smoke. Not being one to miss out on the action, I asked what was going on. They told me that they were looking at one of our tanks that had been hit. Apparently while sweeping an area for IEDs, with a MAP team, one of our Abrams hit an improvised explosive device specifically designed to disable tanks. The tank was completely disabled and was afire. Hundreds of gallons of diesel fuel were burning, along with countless rounds of 120mm high explosive shells and ammunition for the crew served weapons.

The crew of the tank escaped the blast unharmed and was being transported by the MAP team that had accompanied it, in their high-back Humvee. That is until the high back hit a second tank IED. The same type of bomb that had destroyed the most heavily armored tank in the world, had just exploded under a pickup truck. The loss was complete. The Humvee was vaporized. Five Marines were killed, four of them instantly. The corpsman assigned to that MAP team, Robin Thacker, worked feverishly to save the last mortally wounded Marine. He poured every ounce of knowledge and energy that he had into saving that final Marine's life and only stopped when he was pulled from the body by the flight medic from the medevac helicopter.

When I first met Thacker, I found him to be quite intellectual, and. I wondered if he was well suited to be with a Marine Corp infantry unit. After serving with Robin, I can't say that I have ever had a more wrong impression about a person in my entire life. Robin was a true American hero that day, and on several other occasions. If you ever meet him, buy him a beer, because he deserves one. Robin Thacker is a good man.

The same Marines who were watching the blaze with me through the spotting scope were immediately sent out to secure the site. A fellow India Company Marine, and friend of mine, Staff Sergeant Mitchell, was the NCO in charge of the detail. For two days, his Marines sat in defensive positions to prevent mortuary affairs officers from being attacked by insurgents as they collected the thousands of pieces of the dead Marines. And for two nights, Staff Sergeant Mitchell's Marines would mercilessly kill the packs of wild Iraqi dogs who would try to feed on the flesh of our dead friends. I can't think of a worse mission.

Back on River Road at the ambush site, however, the tank that Gunny Mac was behind was doing a fine job of keeping us safe. We had received some small arms fire that had just started to become serious when the tank's coaxial machine gun opened up with a burst of about twenty rounds down the alley, putting a quick halt to the fire we were receiving. Cool. The only thing that would have been cooler was if the main gun was firing at a rapid rate and leveling every building that was within a block of us, but I guess restraint was good, even in the worst of times.

To add to the sense of safety the tank gave me, Cobra attack helicopters were circling overhead. I looked up in time to see rockets pouring from one of the aircraft and slamming to the ground below. Wonderful, I thought. The Cobras were equipped with excellent thermal sights that allowed the gunners to easily locate insurgents who were attacking us. I was sure that the pilot was just picking out targets at will and sending them to their graves.

Gunny Mac gave us a quick briefing. The building to our left was where the ambush started, and it was loaded with bad guys. One team would surround it from the back and one from the front. Then a third would go inside and clear it out. I was assigned to the team that was going to the back and the rest of the Marines from my MAP Team were assigned to enter the building and clear out the insurgents.

I stopped Gunny Mac immediately after he gave me my assignment and asked him to reconsider. I told him that I was on a SWAT team as part of my civilian job as a police officer and had spent years clearing buildings. I thought that anyone there, I probably had the best chance of clearing the building safely. Surprisingly, the Gunny agreed, and put me in charge of a Marine from Communications, Smitty, and a third Marine from my MAP team, Rowland, with the assignment of clearing the building.

When we were receiving our pre-deployment training at Twentynine Palms, a great deal of attention was given to showing all the Marines how to safely clear buildings such as this. All Marines except for the 2/14 Marines, that is. They were driving the trucks that took us to and from the evolutions and had to sit out what was now proving to be a very important class.

The fear that had crippled me in the first few minutes was gone. I was mad, furious, but at the same time happy to be in the fight. Smitty grabbed my flak jacket and yelled, "You just tell us what you want us to do." I loved Smitty. Like me, he was married and a college graduate. He had a common sense about him that was very uncommon in the Marines, and that drew me to him immediately. When he talked, we listened, because what he said was always simple…complete with a thick southern drawl, and it always made sense.

I told Smitty that I would go first, and he would cover me. The next Marine would cover long down the hallway, and the last Marine would cover our rear. It was a formation that I was taught as a police officer for use in school shootings called a "quad formation." After the shootings at Columbine High School, someone had devised this as a quick and easy formation to implement when four people with limited training arrived at a large building where people were inside shooting. Seemed to fit us pretty well.

I kicked open the door to the building and we went inside. The walls on the inside of the building were stacked with UN "Oil for Food" grain sacks, which were set ablaze either by the explosion of the car bomb, or the massive amount of fire that Jeff's MAP poured into the building while repelling the ambush. I don't know why, but as I passed the burning sacks of grain, I started thinking about some of the political aspects of the oil for food program and whether or not it had been effective. But then someone started shooting at me and I started thinking about that instead.

I walked past the flames, and I entered a room to my right. In the corner I saw a chocolate chip camouflage uniform that was issued to Iraqi soldiers. I fired a burst into it immediately. We didn't have Iraqis with us, so any Iraqi soldiers inside would have to be insurgents. As I walked up to my target, I was surprised to find that rather than a person, it was just a uniform. I searched the room and found that the window facing River Road had been broken out, and a step was placed at the bottom to allow someone to climb up and look out. There were spent AK-47 shell casings and blood all over the room, which had obviously been fortified as a firing position.

I took a few deep breaths and reminded myself to make sure of what I was shooting at before I fired. We left the room and continued clearing room after room as we made our way down the hallway. It was exhausting work. In the background I could hear AK-47 fire coming from deeper inside the building. Smitty and I found room after room had been occupied by insurgents but were now empty.

I came to a set of double doors with a pane of glass in each one, the kind that often separated the kitchen from a dining room in a restaurant. As I approached, I saw an Arabic male hiding behind the far door. I blinked in disbelief and reminded myself to make sure of what I was shooting at. As I was trying to figure out what the man was doing, Smitty yelled a warning to me, "Look out!"

I placed the front sight of my rifle dead center of the figure on the other side and fired three, three round bursts from my M-16 through the door into the man.

I was only about five feet away when I shot, with him locked squarely in my sights. But after I fired, I don't remember seeing what happened to him. I don't remember seeing him fall, I don't even remember him lying on the ground as I stepped over him. That was the first person I had ever killed. I honestly didn't even remember shooting the man until about a week later when my MAP team was recounting the incident. Smitty was letting the rest of the team who stayed outside the building know the details of what had happened inside as we were clearing it.

I had heard of certain words or events or objects that triggered people involved in traumatic events. Just a random experience or thought brought back in vivid detail repressed memories as a matter of self-defense. Smitty recounted the scene to the group of Marines explaining, "Doc shot the shit out of this one guy…" And there it was, the entire scene played back in my mind like a movie, like a scene deliberately edited to look like a person having a flash back of a traumatic event. I saw the doors. I saw the man; I saw him reaching. I felt my rifle swinging up from the ready position. I could hear Smitty yelling, "Look Out!" I smelled the gunpowder of the rounds igniting three at a time in burst fashion as I squeezed the trigger, once, twice, and then a final time. I remembered the window shattering, and holes ripped open in the door from the rounds passing through. I remember the man being only inches away on the other side of the door, looking through the same window that had just shattered. Then I remember…nothing. It was like the next five feet of everything I saw had been edited out of my memory. I just can't remember seeing his body, either when I shot him, or when he fell or even when I had to step over him into the hallway.

Had I not been there when Smitty was telling everyone what I had done, I could have gone some time without fully appreciating the fact that I had killed someone. Thank God it happened there, amongst Marines who would understand me, instead of years later when I was opening Christmas presents with my family. Oh, that would have been hell. I remain scared to death of the day that I will remember that man, and his lifeless body, lying there in a pool of blood caused by the wounds from my bullets.

After I shot, the next thing I remember was kicking open the doors and flowing through into a second room. As I passed through those doors, what I saw almost stopped me in my tracks. There were over a hundred o people lining the hallway. Most were women and children; some were very elderly. Inside a room off of the hallway was about a dozen military aged males (MAM).

MAM was a term created by the Marine Corps to describe men between the ages of sixteen and fifty and gave us some certainty as to exactly who was and who was not an insurgent. A fine idea, except that it excluded women who were shooting us, or old men driving car bombs, or children digging holes for IEDs or donkeys with explosives strapped to them walking towards our front gate. Other than that, it was a very helpful term that was eventually expanded to include everyone and everything in the entire country of Iraq.

I was able to mutter "what the fuck" before I realized exactly what we had walked into. A woman from the crowd in the hallway stood up and started yelling at me. She had a white smock on with a red cross on it. She was a nurse. We were in a hospital. The people in the hallway were patients. The men in the room could be the insurgents who had separated themselves from the patients or could be hospital workers. I looked closer and saw that the people in the hallway were all badly injured. Some of them recently, and some of them were chronically infirmed.

The insurgents had set up an ambush using the Haditha Hospital as cover, and what lay in this hallway was the grisly result. I had to choke back vomit for about the third time that night.

I tried to process everything, the mass of wounded and how to help them, the group of men who could be insurgents and what four Marines could do with them. I didn't have the time, ability or supplies to help these people, some of whom were surely wounded from our rounds. We didn't have the Marines to detain, or vehicles to transport the men, and as much as I wanted to, I couldn't just shoot them. As I pondered what to do, someone was nice enough to help me prod me along. A burst of AK-47 fire rang out in a hallway adjacent to the wounded. It was decided, we needed to get more Marines in here.

My team backed out and I radioed what was happening to Captain Moeller. I told him that we were in a hospital, that there were at least a hundred wounded, and that we needed more Marines if we were to continue clearing the building. We left to regroup and formulate a better plan. We left the same way we came in, past the fire which was now spreading to the hallway.

As we left the hospital and ran towards the tank where Gunny Mac had been earlier, I heard a scream. I looked back and saw the communications Marine writhing in pain, yelling for a corpsman. I ran to him and found that his leg was trapped in a hole that was lined with concertina wire. The wire had been placed as a booby trap for Marines who would try to get into the hospital. Somehow, we had all managed to avoid it when we first went inside. Gunny Mac and I were able to free the Marine's leg and pull him to safety. Struggling in pain, the Marine was not able to stand.

"Lovely," I thought to myself. Now amid all this, I had to treat a casualty. I grabbed the Marine and asked him, almost in a whisper, "Are you hurt or injured?" Because if he was injured, I would have to stop my part in this battle and medevac him, but if he was just hurt, then he needed to get up and get back in the war. He looked at me with pure rage in his face, stood up, and grabbed his rifle. Dragging his bloody twisted leg behind him, he told me, "I'm good to go," and took up a defensive position next to the tank.

That was pure guts. No other person in the entire world, other than a United States Marine, would be tough enough to pull himself out of a booby trap and refuse medical attention to allow the rest of his friends to continue with their mission. That Marine's courage still amazes me to this day.

Captain Moeller assembled another team of Marines and told us that we would be going back inside of the hospital, with my team in the lead. Smitty was still with me, along with Rowland, and now Sergeant V.

I started our advance into the building when Captain Moeller stopped me and had me double check that everyone was on the same radio frequency on our personal radios (PRR). I said yes and started into the building. Captain Moller grabbed me immediately and said clearly, "You need to check and make sure that everyone is on the same frequency, Doc." I thought to myself that this was crap, we needed to get back in that building where the shooting was coming from, and we needed to do it immediately. We didn't have time to change radio frequencies, which took forever in the daylight and was next to impossible in the dark. But an order was an order. And this one, as it turned out, was a life saver. I owe Captain Moeller a big thanks for saving our collective asses.

We again entered and were just barely able to make it through the flames of what was now a fully involved structure fire in the near wing of the hospital. We immediately pressed on to where the patients and the suspected insurgents were. Along the way, I heard someone yelling for help. Screams and cries of the wounded and frightened in the hospital had become little more than a distraction by that point, but that cry for help was different. It was in English.

I veered off to the left and approached the room that contained the man yelling for help. I first thought that we had lost a Marine, or that there was a captured civilian in the hospital, but as I carefully moved around the corner, I saw that it was neither. On the floor at the doorway was an insurgent who had been killed by small arms fire. Behind him was the man that was yelling for help. I immediately recognized him as foreign fighter, probably Syrian. It was easy to tell the foreign fighters because they were the only pasty white people with red hair in the entire country, other than the O'Conner brothers in India Company.

I ordered the man in Arabic to show me his hands. His hands were well hidden under an overturned desk, and I was sure that he would either pull out a gun or blow himself up with a grenade. Again, in broken Arabic I told him to put his hands up or I would shoot. I trained my rifle on him and started to squeeze the trigger when an image flashed in my head. It was of a Marine on CNN in the battle of Fallujah who had shot someone in the same circumstances. That Marine would spend the next two years in a court martial answering for his actions. I paused and gave one more order as loudly as I could. Apparently, sensing the seriousness in my voice, the man raised his hands, stood up and walked towards us. Staff Sergeant Brady, who had been responsible for killing the other man in the room, grabbed the Syrian and put zip ties on his hands. Scattered at the Syrian's feet, behind the desk, were two grenades.

Later our Human Exploitation Team (HET) would integrate the Syrian. The HET were Marines responsible for interrogating the suspected insurgents we detained. We joked amongst ourselves that these guys were the ones who were responsible for "shoving bamboo shoots under terrorists' fingernails until they gave up where Osama was hiding." Nothing could be further from the truth. Our HET Marines were consummate professionals who gained a good deal of intelligence by just knowing how to talk to people. Never once was there even a rumor of our HET Marines mistreating a detainee, something that was a credit to those Marines' strength of character.

They found gunpowder residue on his hands and confirmed that he was from Syria. He had participated in the attack and was planning to kill me as I went to help him. To this day I regret not killing him. I am ashamed that I let the prospect of ending up like the Fallujah Marine on CNN endanger my life and the lives of those in my team. And I am utterly terrified that this man may have lived to kill more Marines.

To this day, I struggle with the consequences of my decision not to kill that terrorist. I alternate between feeling like I failed my fellow Marines in not taking the opportunity to kill someone trying to kill us; after all, that was exactly what we were over there to do. I also felt like life is something so precious, something so final that once taken, it cannot be undone thus it is always better to choose not to act if ever there is doubt. I guess that is how I feel now, most of the time, which is a fine and decent way for any human being to feel about his fellow man. But when fighting a war, and our decisions, right or wrong, have to be made instantly, there isn't time to have an internal, moral debate about death.

It's funny how a man can believe passionately in both sides of the same argument.

On that night, as Staff Sergeant Brady was taking the Syrian out of the hospital, I prayed softly and sincerely, "God, please give me the wisdom to know who to kill, and the courage to do it swiftly." More than anything, I wanted to leave that night without having killed anyone who was truly innocent. Being engaged in a fire fight inside of a hospital made that a very distinct possibility.

After the insurgent was led away, my team again pushed to the hallway where the patients and suspected insurgents were. When we arrived, of course, the dozen men that I had encountered earlier were long gone. Left behind were the sick and dying Iraqis whose only concern before tonight had been getting medical attention for themselves and their families.

I paused briefly at a little girl who was in her mother's arms. She had a feeding tube, which had been disconnected when she was forced to take cover in the hallway. Her face was pale, and waxen, her mouth agape. Her eyes were fixed at the ceiling. She was breathing…barely. I looked at her emaciated frame and knew immediately that there was nothing I could do to save her; she would die soon. Her mother held her in her arms and looked through me as if I wasn't there. Like this was just another day like the ten thousand she had before. That this was the way things happened in Iraq. Her face was so completely void of emotion, of hope. It was the incipient face of a martyr who had witnessed her children dying, her hospital blowing up, and perhaps Americans accidentally shooting her husband. I had hoped to come to this country help this woman and thousands just like her. Instead, the insurgents put me in a position where it was my bullets flying down that hallway and ripping apart the flesh of little girls like that dying one.

I hate them. I hate people so weak and evil that their only purpose in life is to kill and destroy and take from us every advance that we make towards peace. Those people are weak. They live by creating fear and death. They think that they are brave because they claim to not fear death and revel in the prospects of what lies beyond. That is cowardice. True courage is living well, doing the right thing, caring for others and standing up for what is right and defending those who hold the same values. The bravest men live their convictions every day. Those insurgents, who picked that hospital and those people to hide behind, were not men; they were animals.

Knowing that there was still a war going on past this hallway, I pressed through the sea of souls with my team. We continued down a long hallway and again started clearing room by room.

I entered a room to my right. It was a large, open room with frames where picture windows had been before they had been blown out. This had been the front lobby of the hospital that overlooked a small parking lot. As I made my way to the center of the lobby, I saw a flash from a shadow beyond the parking lot, a muzzle flash. "Sniper," I thought to myself. From the instant I saw that muzzle flash, everything was in slow motion. I know that is such a cliché, but that is exactly how it played out. The flash looked long and slow, like someone had lit a cigarette lighter. I clearly heard the word "sniper" in my head, and I remember telling myself to get down. It was as if I could feel the message being passed from my brain, down my nerves to my legs. I could feel my knees bending a millimeter at a time when what felt like a sledgehammer hit me in the head. The sound was deafening, like an explosion. I jerked back into a desk and slumped down to the floor. "He got me," was the next thought that went through my head.

My ears were burning, and I was all but completely deaf. My head felt like a hand grenade had gone off inside of it. I shook off the cobwebs and before I could even process what had happened, I was lifting my rifle over the windowsill and firing blindly. I let fly an entire magazine in the direction of where the shot came from, firing on my rifle's "burst" setting, which fired three rounds on automatic with each pull of the trigger. With each trigger pull, it seemed to take forever for one round to fire then the next and the next. I remember thinking that the chances of me hitting the man that shot me were overwhelmingly bad. I had hoped that just by firing back, he would know I was still in the fight and maybe be scared off. I hoped. And I prayed.

Right after I thought that reason took over, and life seemed to play at normal speed. The insurgents we were fighting were no strangers to being shot at and wild rounds fired lazily in their general direction were not likely to even concern them.

I dropped the empty magazine to the floor and already had the second one in my hand. I reloaded in one fluid motion, just as I had been taught. In my head I heard my instructor from SWAT school years before yelling about the importance of hitting what I was shooting at in a gunfight because, "ONLY HITS COUNT!" I was in it now, fighting for my life, and panicked rounds fired God knows where were not going to get me through this. I popped up in the window with my rifle tight into my shoulder, searching for that son-of-a-bitch that took my hearing.

I saw him, half slinking across the parking lot. I locked my front sight post on him as he moved and slowly squeezed the trigger. I fired one burst at him and then another. I was dead on, I knew I had hit him, I was sure of it. It was just like shooting on the rifle range. I just knew that I had fired the round perfectly and I didn't need someone to tell me that I had hit the target; I just knew it. It felt right.

But after my last burst, a hail of bullets again forced me to the floor. Sergeant V was pinned around the corner in the hallway, unable to come to my aid. The rounds kept coming in, so I yelled to Sergeant V that I needed cover to get out of here. Twice he tried to clear the corner and place suppressing fire only to be driven back. Finally, I yelled to him, "You just start shooting and I'm gonna start running." And that's what we did. Sergeant V shot and I ran…well crawled…as fast and low as I could.

We laid on the floor of the hallway trying to take what cover we could. I barely heard the radio crackle to life over the ringing in my ear. The voice on the other end yelled out a panicked, "Stop shooting at us!"

What the hell? "Doc, stop shooting at us!" It was Wyzorik, one of the gunners from my Map team. I yelled back, "I'm not shooting at you, you're shooting at me!" A short and intense argument ensued about who was shooting at who and why, but the order was given to cease fire. Thank God.

Outside of the hospital, the vehicles from my MAP Team were in a blocking position on the other side of the courtyard about a hundred yards away. The insurgent sniper had maneuvered to a space between the building and my MAP Team. When he saw me inside, he took a shot. When I fired my first magazine wildly over the windowsill, I was in fact sending rounds past the sniper and at my MAP Team. Thinking that they were being fired at from the building, they returned fire.

Enter Captain Moeller and his earlier order for me to coordinate the radio frequencies. Before he told me to do that, Wyzorik and I had been on different channels. If we remained on different channels when we were shooting at each other, the order to cease fire might not have been given until it was too late. Thank you again, Captain Moeller.

The four of us rallied and continued past the doorway, making sure that all who followed knew its peril. We were met by Captain Moeller, Major Blanchard, and the Battalion Executive Officer, or XO. The XO had a pair of NVGs in his hand and held them up as if to say, "whose are these?" Heck, he might have said that, but my hearing was so shot I could only see his lips moving. I reached up to check my gear and found mine were gone. I raised my hand, and he tossed them to me, complete with a bullet hole directly through the middle. I had been shot squarely in the NVGs and between that heavenly piece of equipment and my Kevlar, I was spared. No blood, no brains oozing out, just a blown-out eardrum and a bit of a concussion. Wow. I felt like I had won the lottery.

When we were debriefing the incident later, we discussed just how two groups of Marines found themselves shooting at each other, when one of our snipers described a tactic on how to incite a "blue on blue" (or friendly fire) confrontation. The sniper places himself, with great stealth, between two enemy forces in close proximity. Then, he engages a target with one shot. The returning fire from his target's supporting forces would naturally be fired directly into his friends, which are located squarely behind the sniper. The sniper would then seek cover or displace after firing, while the two same side forces tore each other up exchanging fire. It was a classic sniper tactic. It was brilliant in this situation, and it worked marvelously that night…for the bad guys.

I tend to think that night was a fluke, if it were not for another occasion several months later. We were at a firm base in the city of Hit, when a patrol several blocks out came under fire. The report came over the radio that the patrol was in contact and returning fire. At the same time, the Iraqi Army sentries on the roof reported that they were in contact and returning fire. Everyone at the base responded, and a good volume of fire was returned in the direction from where we assumed the enemy was positioned…which was the location where our patrol was pinned down. We were again shooting at each other. Luckily, we figured this out before anyone was shot and ceased fire. One insurgent had snuck in between our base and the patrol, and with a few well-placed shots, started a war between Marines.

The insurgents were taking advantage of our biggest strength, an overwhelming volume of fire, and turning it against us. It was very smart; I have to give them that.

After the sniper was dealt with, my team and the group of assembled officers pushed into the final portion of the hospital. I was approached by a nurse who was yelling wildly at me in Arabic. Major Blanchard instructed me to tell her that the hospital was on fire and that she would need to make provisions to evacuate the injured.

I looked squarely at the nurse and said slowly and loudly…in English: "Your hospital is on fire. You have to go." I added some hand motions to try and simulate the words "fire" and "go" in sign language. I only spoke two words of Arabic. I looked like an idiot and the nurse looked even more confused than before.

"Jesus, Doc, I could have done that," the major remarked, obviously disgusted at my inability to speak Arabic.

While my animated exchange took place with the nurse, the chief physician of the hospital arrived. Thankfully, he spoke passable English. At once he begged us to stop firing into his hospital, saying that he had many wounded from the exchange of gunfire. I assured him that there would be no more rounds fired into his hospital, and that we would keep him and his patients safe. I said this sincerely, with a feeling of pride that I was able to protect people who needed my help. That's why I had gone there and I felt emboldened that I was now in a position to be the protector instead of the infidel.

That feeling didn't last. The doctor crushed my pride by instantly accusing the Marines of attacking his hospital, without provocation, and killing innocent civilians in the process.

I stood speechless for what seemed like five minutes. I tried to process the complete dishonesty from the man who stood in front of me. Lying, unfortunately, is practiced by many Iraqis, but this man had elevated it to an art.

At the same time, I was trying to wrap my head around his accusations, my hands were balling tightly into fists. I was grinding my teeth and trying to come up with something productive to say to improve our situation, instead of just screaming at him. That failed to work, because screaming at him was exactly what I did.

"We got attacked from inside your hospital. I shot a man inside your hospital. Someone shot me in the fucking head while I was inside of YOUR hospital. We captured an insurgent who was trying to kill us just down the hall!"

The doctor was a bit shocked at my tone, and probably quite intimidated by a giant figure with a machine gun walking towards him and yelling. Whatever the case, his demeanor turned apologetic from accusatory and said, "They are gone now. Please stop shooting."

I believed him. The insurgents were gone from the hospital. The small arms fire and occasional explosions that played constantly in the background had all but stopped. The fight was over. Even though he was cultured in deception, this man was still a doctor, and appeared to have at least some semblance of caring for his patients.

Satisfied that the hospital was cleared, the Battalion XO, along with Captain Moeller and Major Blanchard started to plan our next move, and how to get out of the hospital. The door we entered through, and more accurately the entire wing that surrounded it, was now ablaze, and there would be no leaving the way we came. The fire was spreading to the rest of the hospital quite rapidly.

I told the good doctor that he needed to move the patients because the hospital had now become untenable and would soon be lost. The doctor organized some of his nurses and others standing around and started moving the patients in the immediate area out of the building. Marines from our group helped the evacuation while an egress plan was formed.

Major Blanchard and Captain Moeller looked around the hospital, and at the pile of wounded Iraqis littering the hallway. "Doc…see what you can do." Captain Moeller's order wasn't barked with the urgency of a life and death command given by an officer to his troops, it was sort of hopelessly whispered by a man who realized that many of these Iraqis would die despite our presence. And that a few of them would die because of it.

I started examining the wounded, providing what care I could, and trying to triage them the best I knew how. For a minute, I thought that this was the new plan. We would help evacuate the hospital, I would help care for the injured and maybe even coordinate a medevac to a military hospital for the most seriously wounded. Heck, we were here to help, and these people obviously needed our help.

I moved frantically from one patient to the next until I heard Captain Moeller yell to me "Doc, let's go." Our group of Marines was formed up and on the move out of the hospital. We were leaving…just like that.

I wanted to object. I wanted to say that we should help these people, we could be the good guys, we could keep the moral high ground if we just stayed and helped. I wanted to tell my officers this…but I didn't. Instead, I dropped the bloody hand of the person I was treating and fell into the rear of the formation. For the second time that night, I was ashamed because of my inaction. I had missed an obvious opportunity to do the right thing.

The hospital would burn to the ground that night, leaving a city of over a hundred thousand with no medical care. Back in the States, the student paper for Kent State University (my alma mater) would denounce our role in viciously attacking a civilian hospital.

To this day I wish that we could have stayed and helped those people. That night, and even today, I realize why we didn't. We only had half enough Marines to protect ourselves, let alone a parking lot full of wounded. We didn't have the logistical assets to move all the patients, and our hospitals were filled to the brim with treating wounded and dying Marines.

Likewise, it was clear the insurgents were permitted to occupy the hospital as a base for attacking US personnel. That was my first lesson in the Iraqi medical system, which I learned was completely anti-US. When I went to Iraq, I thought that I would be working beside the local medical providers, helping them, and teaching them. My experience provided otherwise: much like the local mosques, hospitals and clinics were bastions of insurgent activity. We didn't want to attack them and were prohibited by the rules of engagement even if we wanted to. I think that if anyone knew that the building we were being attacked from was a hospital, we would never have entered it. It breaks my heart that someone would hide behind religion and the infirmed. I find it difficult to understand people who could do that.

We left the hospital in a tactical column and fell in on defensive positions to guard against a counterattack while we were working on our exit strategy. I remember climbing a pile of rocks to the west and laying down next to Gunny McIntyre. "Doc, move your guys up on line with us, and tell them no one comes through," he directed. I yelled to Smitty and the rest of the team to get on line, and they fell in the best they could.

And so, we laid there, and waited for either a counterattack, or the order to pull out. All our dead and wounded were recovered, we had retrieved our equipment, and our vehicles, that were not burnt to the ground, were up and running. Laying in the still of the night was the first break that I had since arriving. I looked at my watch to see we had been fighting there for four hours. All at once it seemed like minutes and days. Pain started to set in my back and head from where I had been shot, I was bleeding from my foot where I had been cut by the concertina wire from the booby trap that crippled the communications Marine, my hand had shrapnel in it and was burnt from traversing the fire, and my hearing was now almost nonexistent. I looked to my left and right at the line of Marines and no one looked in any better shape.

As I sat there, fear set in again. I prayed to God to take us from this place, to let us leave and not have to spend the night. I wanted this nightmare to be over. I started to feel my emotions taking over, but I forced myself to snap out of it myself almost immediately.

Then the order came to move out. My MAP team mounted its vehicles and followed in line behind a tank. The danger of a second attack or IED was still very real, so officers were on the radio giving direct security commands to all the vehicles. I had the radio to my ear and was not even able to distinguish that someone was calling me. Captain Moeller recognized that deaf, I was not going to be of any use as a radio operator. He took the headset from me and let me just stare out the window.

We arrived back at the dam and drove directly down to the bunker. No one spoke as we dismounted. I crawled out of my Humvee with some assistance from the administration captain, who helped carry me to the BAS for treatment.

As I walked in, I could see Chief Schaser, Captain Gratton, and Senior Chief Naida, all of them covered in blood. They looked as bad as I felt. I could tell they had been crying. I imagined that they had as rough a stretch here as we did at the hospital. I looked around the treatment tables at the bloodstained floor and piles of bloody bandages. It must have been hell for them to sew together their friends for the past four hours.

Chief Schaser grabbed me and gave me a hug as he was crying. Before Iraq, I had not been a fan of men crying and hugging, but this would be the first of about a thousand times the Chief and I would comfort each other in such a way, and I will forever be grateful for the experience.

My wounds were treated as the Chief recounted their night. It had taken about fifteen minutes to unload the seven-ton when it arrived, because the bodies of the dead and wounded were packed so tightly. The treatment of many wounded Marines was delayed because they were trapped under the bodies of their deceased comrades. Four Marines had been killed, a dozen or so wounded, all severely, with the issue of some of their survival still in doubt.

Staff Sergeant Priestly, the battalion armorer and paramedic from Cleveland, would fight a battle against the amputation of his legs for the next year. That hurt. He was a good man and a real asset to have in the medical department when we needed extra help. He had volunteered to help Jeff's MAP out on this mission, because he wanted, like everyone, to do his part.

After a good accounting of the night at the BAS, no one had mentioned Jeff. Senior Chief Naida, still crying and barely talking above a whisper, asked to see me outside. He asked in a way that someone a thousand miles away could tell what was coming next. I didn't move from my seat. I didn't go outside. I just looked up at Senior Chief Naida and said matter-of-factly "It's Jeff." Senior Chief Naida dropped his head, half in a nod, half in anguish, confirming what I already knew in my heart to be true.

I fell to the ground screaming, crying, and hurting in a way I had never experienced before. All sense of reality left me, and I just laid there crying until I couldn't cry any more. For a moment, I stopped. A thought flashed into my head, and I asked hopefully, "Is he dead or just wounded?" No one answered, they just looked at me with a mix of pity and emptiness that there was nothing they could do to console me, they could only stand there and let me cry myself out.

When Jeff and I first arrived at the dam, we were organizing our equipment in the BAS. Jeff pulled out an American flag that he had brought with him and hung it over the doorway. "The last thing I want wounded Marines to see when they leave here is that flag. I want them to know why they are here and that the whole country is behind them." That was the last thing the Marines who were wounded tonight saw before they were loaded onto helicopters. As I sat there on the floor of the BAS, Chief Schaser stood on a chair, and pulled the flag down from where it was hung. The Chief walked over and wrapped it around me.

"Jeff would want you to have this," he said as he embraced me. Of the things I carried through the war, that flag, and Chief Schaser's friendship, remain two of the most important. Brett Schaser is a good man.

Captain Gratton walked in a while later. He said Jeff was still at the BAS and asked if I wanted to say goodbye. I told him no, that I didn't want to remember him like that.

"I know; I couldn't bring myself to do it either," Gratton told me softly. Those words came reluctantly out of the throat of a man who had seen more than enough dead people in his lifetime.

There would be a debriefing in the Operations room for everyone involved in the attack in Haditha. Captain Gratton spoke up, his tone changing enough to let everyone, including Doc Cleveland, know that they needed to get back to the business of fighting a war. I was expected to attend the briefing and report what had happened in the hospital. I straightened up the best I could, stopped crying, and started mentally preparing for giving my report.

The war had a funny way of helping me cope with loss. I never had time to grieve. After every Marine that was killed or wounded, there was still a job to do. And if we really wanted to honor the memory of those we grieved for, the best way to do so was by getting back into the fight and doing our jobs. If we were lucky and made it home from this mess, we should have about sixty years to pour over all our blown-up friends. But for now, it was back to work.

So, I walked out of the BAS, feeling as if I was only part of the man that I had been just hours ago, and knowing that there was still plenty of work to be done. Looking back, I can't say as if I have been changed more by the passing of any five years of my life than I had been in those four hours.

I arrived and sat among the twenty or so Marines in the Operations room. Not a word was said by any of the warriors, but the looks on their faces all told the same story of sorrow and exhaustion.

Major White, the Battalion Operations Officer, stood in front of the crowd to conduct the briefing. Obviously as emotional as the rest of us, his first words were directed to us at a yell: "We did not just get our asses kicked! Do you hear me? We did not just get our asses kicked!!!"

I was shocked at what he was saying, because I can say for sure that we were all feeling like we had just lost a fight. I wanted to tell the Major that I had my ass kicked before and it felt exactly like this. We spent the next hour or so rehashing exactly what had happened. I gave my ten-minute dissertation about my part inside the hospital.

Everyone seemed particularly interested in the fortifications for firing positions, and the detainee we had taken. There was the expected discussion about the blue on blue, but it was all sorted out in short fashion.

I went through the rest of the war feeling that Major White's speech was made specifically to cover up the fact that we had taken a beating. Thinking back though, he was exactly right. We had taken on a numerically superior force that had a wonderful tactical and strategic advantage, and we did so with a fraction of the normal forces that we would have, most of whom were not infantrymen. We had not been beaten. This was war. Every war has its losses, and in the end, we just hope that we gave a little better than we got. We did that and then some. Major White wasn't trying to tell us anything other than that.

For the first time I heard the whole story of what happened to Jeff's MAP that night. As they traveled down River Road, they stopped in front of the hospital because a car blocked their path. The Marines inside the vehicles dismounted for security as the column turned around. As Jeff got out to help, Corporal Schueler told him to "stay in the truck," to which Jeff promptly responded, "Screw you!" After the Marines were out on foot, a car loaded with hundreds of pounds of explosives slammed into the side of the Humvee killing the Marines inside, one of whom remained trapped until freed by First Sergeant Sauers.

Two other Marines died instantly, as insurgents from the hospital ambushed en masse. Jeff fell immediately, struck by shrapnel from the blast. A Marine came to his aid and tried to treat his wounds, but he died in his arms. He died right there, in the middle of that street, River Road, Haditha, Al-Anbar Province, Iraq.

I can't understate the heroism that took place that night. Then Lance Corporal Todd Corbin, a thirty-year-old sheriff's deputy from the Huron County Sheriff's Office, would be awarded the Navy Cross for his heroic actions in directing Jeff's MAP team in repelling the ambush, evacuating the dead and wounded, and rendering all our most sensitive communications equipment inoperable, as not to allow it to fall into insurgent hands. Many other Marines were heroes that day as well, Corporal Shuler, (who would be awarded a Silver Star and later become commissioned as an active duty Marine Officer) First Sergeant Sauers, the communications Marine who refused evacuation, Smitty, Captain Moeller, Sergeant. V…Jeff…the list goes on. It is humbling to have been a part of what happened that day, and I am proud to even be mentioned as part of what those heroes have done. God bless them all.

I limped back down the dam to our bunker, alone. I remember thinking "What now?" I had no idea what was to come, how, or if, I would get past this. I was so alone. I climbed into my bunk, certain that I would be refused sleep by the memories of what I had just done. Thankfully, exhaustion took over and I slept so deeply, that I was unaffected by my dreams.

I woke the next day, still in a haze from what had happened. I limped back to the BAS for my wounds to be rechecked. It was morning in Iraq. Back home, everyone would just be falling asleep. In about eight hours, when Jeff's wife woke, she would get a knock on her door from a Navy Chaplain and several Staff NCOs from the local Marine Reserve Unit in Long Island where they lived. Maria

Wiener, proud mother of two, would forever remember Mother's Day as the day the father of her children was taken from her, four thousand miles away.

Never would she hold him again, never would they dance at a country bar. The girls would both be given away at their weddings by another man. The girls would be forever robbed of their father. Eventually, they too would become mothers and have children who would also be robbed of their grandfather.

Most heart breaking, the girls were just young enough that the memories of their wonderful father would fade, to be replaced by only stories of his love, passed on by their mother, who would never allow him to be forgotten.

Why? Why Jeff? I kept asking myself that over and over. Not having an answer, I started asking God. Why Jeff? I wanted, no, I needed to know: why Jeff? More importantly I needed to know why not me. Why was I to be given the opportunity to do all the things that had been taken from Jeff and his family? If the round that hit me had been an inch lower, I would also be dead. If the shrapnel that hit Jeff was an inch to the side, he would still be alive. What gave me the right to live through this war when Jeff did not?

And then it hit me. I hadn't lived through the war. I had lived through four hours of one engagement, and I had five months to go.

The same day, a Lima Company Amtrak had hit a mine near the Syrian border, killing fourteen Marines. We had lost twenty Marines from our battalion in the first few months of our tour, two to three times what a typical battalion would suffer in an entire tour, and our operations tempo was just increasing in speed and intensity.

We were taking significantly more contact and there was no sign of it stopping. No, I hadn't survived the war, and quite frankly, the issue of whether I would or not was seriously in doubt.

I called Louise the next day, when the phone center was re-opened. She lived every minute in a general state of fear that something horrid would happen to me. When I told her that Jeff had been killed, her fears moved into the category of certainties. She cried. I imagined her collapsed on the floor. She cried for Jeff, for Maria, for her girls and because she was now certain that she too would be receiving a knock on her door from some nameless Navy Chaplain. I didn't even try to comfort her. There was nothing I could say that would help.

I called Maria next. We talked for about ten minutes. I said only two words. "I'm sorry." I said it over and over again. I cried, I wailed. "I'm sorry…I'm so, so sorry." I was sorry I took her husband from her. I was sorry I talked Jeff into deploying with India Company. I was sorry we didn't stay at Al-Asad. I was sorry I didn't get there in time to save him. I was sorry she was a widow. I was sorry her children were fatherless. I was sorry I was not there to comfort her. I was sorry I lived, and Jeff did not.

I would spend the next week recovering. The shrapnel wounds and burns on my hand were minor and would heal quickly. My hearing was returning to as good as it would get, and my headache was slowly leaving me. I had developed a staph infection where the razor wire from the booby trap had cut me. It was moving up my leg nicely and required me to take some time lying around.

That sure was a blessing. I was having a real time trying to wrap myself around all that had happened. My best friend was dead, and I had been shot in the head. This was not supposed to happen.

I missed Jeff terribly. I felt so empty without him here. It was so unfair. He was a wonderful father of two gorgeous little girls. He was the type of father that made you ashamed that you didn't treat your children that well. He lived for his wife, Maria. Never once, when he was around the legendary womanizing that Marines do, was Jeff even tempted. He was just so happy with her. His patriotism and purpose were unwavering and had guided me through these past months. Now I was alone. What the hell was I supposed to do?

Probably worse than the feeling of loneliness was the stark reality that I may not live to see my family again. I may never meet my son. I might not walk my daughter down the aisle at her wedding. Dozens of times before, as a police officer and firefighter, I could have easily been killed. The thought of dying was always in the back of my mind, but was usually an afterthought, something that happened to other people and definitely not something that happened to me.

I had an arrogant feeling about living. I felt that my survival had thus far depended on how capable I was at my job. Jeff's death showed that someone who was even more capable than I was, could die just as easily. I was coming to understand that who lived and died had little to do with how proficient a Marine one was and had more to do with happenstance. It was a game of inches, and my role was entirely insignificant in the outcome.

As I thought about Jeff, I marveled at how wonderful it was that he was able to see his wife one last time in Las Vegas before we left. Jeff had tossed the idea around of whether Maria should meet him for the few days leave that we had before shipping out. He didn't want to put her through yet another tearful goodbye. In fact, he had all but made his mind up that the trauma was not worth the benefit, and that Maria would not come to Las Vegas.

For whatever reason, logic was defied, and she did come. Jeff and Maria passed the time doing, what Jeff described as, "what a husband and wife do." Fate would let them have a precious few last days together.

When I look back at a set of events, after knowing the outcome, I can see how things happened for a reason. Like Jeff seeing his wife in Las Vegas, or his brother Josh for the last time in Al-Asad or telling me "It will be alright," in the hallway of the dam as we spoke for the last time.

Jeff had a second brother, David, the youngest. Jeff always worried about him. Not because David was prone to any particular trouble, but Jeff was the oldest and he felt it was his job to make sure that his little brother was taken care of. David had enlisted in the Marines about the same time Jeff deployed. Besides his kids, it was all Jeff could talk about, how David was in Parris Island, what he was doing, how he couldn't wait to see him in Camp Lejeune when we returned.

David was at the School of Infantry when we arrived back in Camp Lejeune. I contacted the officer in charge of David's training and arranged to spring him from class and allow him to attend the formal memorial service that the 2nd Marine Division was having for the forty-eight fallen members of the battalion.

When I picked him up, all I could do was cry and apologize for not bringing Jeff home. That came as no surprise to me, I knew I would break down. What surprised me was how little I cried. I thought for sure I would wail like I did when Senior Chief Naida told me Jeff had died, but I didn't. I had been crying over Jeff's death for about six months by that point. Add to that the hundred other mangled bodies that were piled onto my conscience since Mother's Day and that made me officially emotionally broken. I was spent. I was done. If every Marine on the whole base suddenly dropped dead, I would still have had only about three tears left.

I tried to tell David about what his brother was like, the things he did, and what happened when he died. I was doing my best, but I just didn't seem to be doing the moment justice. David and I arrived with the rest of the battalion at the memorial service. It was held in a grass amphitheater overlooking Onslow Bay. Arraigned in a semi-circle, at the base of the amphitheater, were forty-eight rifles, each held upright by their bayonets which had been driven into the ground. Atop each rifle was a helmet, at the base a pair of boots, and hanging on the grip of the rifle, the dog tags of the fallen Marine. The semi-circle was well over two hundred feet long. Never in all my life have I seen anything like it. So many departed souls, loved by so many men, gathered in one place.

The service was what I expected a Marine Corps memorial to be. The ceremony was planned meticulously and performed flawlessly, with all the honor and respect that the Marine Corps showed to each of his fallen for over two hundred and forty-five years.

After the ceremony, I introduced David to Todd Corbin, Jeff Schuler, and some of the other members of Jeff's MAP. They talked for about an hour. The MAP team treated David as if he were a long lost member of their family. They hugged him, they cried, they loved him instantly. They told stories of how Jeff lived, and they told the story of how Jeff died. In the end, I am not sure who benefited more from that meeting, David, or the Marines of K-Bar-7.

It was nice that God let Jeff say his goodbyes, even if Jeff didn't know he was doing it. And when Jeff was unable to say goodbye to David, it was providence that put David in the middle of a dozen

Marines who could say goodbye on Jeff's behalf. Everything happens for a reason.

A few days after Mother's Day, my father was in a serious car accident, and was in intensive care. I received a Red Cross message about the crash, and I called Louise to hear the details. He had pulled out in front of someone and had been unconscious directly after the accident. He had serious chest injuries which necessitated a chest tube. His condition appeared to be serious, but stable.

Besides his health, of equal concern was who would take care of his farm. Louise, with two small children, was in no position to do so, and it appeared that someone would have to be hired for some time. I spoke with Captain Gratton about the matter and asked for emergency leave to return home and make arrangements for the farm and see to my father's health.

As I awaited a decision on my request, Jeff and Maria's final days together kept playing through my mind. Was my dad's accident all part of some grand plan to allow me to go home once more to meet my son and say my final goodbyes? If I did get to go home for a week or so, would my returning mean my death? I started hoping that my leave would not be granted, that surely, since God had given Jeff a precious few days with his wife, he would do the same for me. So, if my trip home was to be my last, then staying would mean that I would survive. As I look back on my whole internal debate about whether I would live or die, I have to laugh. How silly was it that I actually sat down and planned out what would happen if I went home or not? As if any of that mattered.

In those few days of waiting, I had what could only be described as a moment of clarity. I was running on the dam, as I did every day. It was dusk and the sun was setting over the western part of the Euphrates reservoir. The sunset always was a deep purple red that reflected off the crystalline water. The smell of the water, the breeze off the lake, and the setting sun made me think about every night when I was back home, looking over Lake Erie. That was part of the reason I ran every chance I had. For that brief period, right before night fell, I was back home.

By the time I finished running, the sun would have set, and I would be left standing under more stars than anyone could possibly imagine. That was when, no matter how much I was reminded of home, I was unmistakably in the desert. Nowhere on earth could I have had a more complete celestial view than in the sands of Iraq. It was humbling. After my run, every night under the stars, I would pray.

The stars, this place, made me feel so close to God, closer than I had ever felt before. Like being in Mesopotamia, looking at the same stars the Hebrews had and standing next to the Euphrates River somehow made the distance between heaven and earth not so far.

Every night I asked God to give me the wisdom to make the choices he would have me make, the courage to take his path, and the ability to accomplish whatever task was set before me. That night, I struggled with the issue of my life or prospective death during every step of my five miles.

I started off my prayer with the words, "God, please take care of me." I heard God's response not in my mind, but in my ears as if the words were said out loud, "Haven't I always taken care of you?" As I stood under those stars, a feeling of peace and calm came over me and made me feel safer than I had ever had in my own backyard.

God would take care of me, no matter what. Whether I lived or died was of little consequence. I had lived a good life, married a wonderful woman who gave birth to two lovely children. If I were to live, surely, he would guide my life with them when I returned, and if I were to die, he would likewise guide their lives in my place. God had a plan for me, and live or die, I believed that God would have no harm come to my family. I had led a good life, and if this was the end of it so be it.

And that was how I lived the rest of the deployment. It was no longer about me, about my life. Living or dying no longer mattered. There were many times when I was scared, but each time I found myself afraid, I found comfort in the fact that all of this was part of the plan, that happened had already been decided. What mattered was how I acted in that moment. What mattered was the mission and my Marines.

The notion that war creates infallible spirituality in those fighting in it is a well-known cliché. I don't know if I necessarily agree with the part about war "creating" spirituality in a man, but it definitely awakens it. I've always been a religious person. Catholic school, altar boy, church on Sunday…the whole bit. I can say though that there were certain times in my life that I have felt much closer to God than others. Many of those, the birth of my daughter, my wedding, the death of my mother-in-law: all came without war. But a surprising number, church in a sandstorm at Twentynine Palms, Ash Wednesday service soaked to the bone in camouflage face paint, the birth of my son, and Mass in a hydroelectric dam in Iraq with just the priest and I: all came as a part of my military experience.

When I look for the common denominator in each of those experiences, I think it was powerlessness. To be faced with an event so awesome; death, birth, love…and to know that the effect that I have on the outcome was insignificant, drives me to the one idea who holds power over such things.

I think that war, at times, is the personification of powerlessness, Mother's Day certainly taught me that, and it was in those times that I reached out for help.

Each day that passed after that was better. I had been given a bit of time to heal, to recharge my batteries, and to get right with myself, God, and all that had happened. I wasn't better, and would never be the same, but I was coming to accept that this war was going to transform me. I couldn't control what would come, but with some introspection, I could understand its effect on me. I also felt as though I had been spared for a reason. That there would be more to do throughout this war, and that when the time had come; God would put me where he needed me.

About a week after Mother's Day, we had a memorial service at the dam for the K-Bar 7 Marines, and the Marines from Lima Company who were killed in the Amtrak explosion. Twenty rifles were placed, bayonet down in a sandbag, the helmet of their owner set on top of the stock, and the dog tags of the fallen Marine draped over the handguard. One by one, Marines passed and touched the helmets, saying their final goodbyes to their friends. Someone was chosen to memorialize each of the fallen, I was asked to speak about Jeff. This is what I said:

> *"I have neither the words, nor the ability to tell you about my best friend, HM2 Jeff Wiener. Even though it took years to know him like I did, anyone who met him for just a few minutes was immediately struck by his strength of character. Jeff was just physically imposing enough to make you fear mispronouncing his name. And invariably when people did, usually pronouncing it WINER, he would announce, in a voice loud enough for everyone to hear, and in an unmistakable Long Island accent, 'It's WEENER, like the balls!'*
>
> *You see, Jeff was proud of who he was, of every aspect of his life from his family to his name, to his heritage. But what he was most fiercely proud of was his service with the United States Marine Corps. When we met initially at Camp Geiger, there was one thing that drew us together like brothers. We both thought the Navy was gay, and we wanted to serve with the Marines. Not just the Marines, but with a line company. Jeff was from a Marine Corps Family, his brother who is a Sergeant with 1/7 served the last year in Al-Qaim. It had been over a year since Jeff had seen him. As fate would have it, Jeff was able to see his brother for the first and last time in Al-Asad, as he was leaving the country and Jeff was arriving. Jeff's other brother is in Paris Island right now and Jeff had planned on visiting him at SOI when we arrived at Camp Lejeune.*
>
> *Jeff was more of a Marine than most Marines I know. That is the highest compliment any Marine can ever give to his doc, and one that is handed out only sparingly. Jeff earned that status among his Marines, and only a handful of corpsmen in this battalion can say the same.*
>
> *I have met many corpsmen in my career and Jeff is easily in the top five best I have ever met. Just a quick note about his ability. We had attended a Special Operations Medic class in Camp Lejeune prior to being deployed. The class was taught by a private company who hired ex-recon and SEAL corpsman for missions with the FBI. The instructor of the class was so impressed with Jeff during several evolutions that he offered him a job on the spot. Jeff was truly the best of the best.*
>
> *He was a warrior. His only drive, every day, was to go to war with his Marines. Jeff was always looking for a hot mission, telling me that he*

couldn't live with himself if he just laid back in the rear for the whole war. Jeff never laid back, he never took a play off, he never gave up. Before being deployed, Jeff tore his ACL, and suffered what his doctor described as a "phenomenal hernia." Any one of these injuries would have caused him not to be deployed, but he never spoke up, not even when he was taking Motrin by the hand full just to go on a patrol. He died a warrior's death; he is a true American hero.

If there is a person who valued service to his country more highly than Jeff did, I have yet to meet him. A paramedic in New York City prior to joining the military, Jeff lost several friends in the attack on the World Trade Center. Many times, when I would find myself questioning why I am here, I could just look to Jeff, and I would know. Jeff had an unwavering sense of purpose. He knew why he was here. He was here to make the world just a little safer for his daughters to grow up in. He was here because his country called him. He was here because some of his friends no longer were. Of all the things that Jeff gave me in his life, that sense of purpose is the most valuable. I no longer question why I am here; I know why. I am here to make the world a little safer for my kids, I am here because my country called me. I am here because my best friend no longer is. God Bless you Jeff.

Semper Fi."

Jeffrey Lewis Wiener was a good man.

PHOTOS

Part of India Company, 1st Platoon along with their Iraqi Army Counterparts, taken in the City of Hit. Andy Lenz is the tall Marine standing in the center rear

This is what remains of the AMTRACK that struck a mine during the assault on the City of Haditha during operation New Market.

Edward Barnett and Robin Thacker standing on top of the Haditha Dam.

Brian Hill at the FOB 1 aid station in Hit after a stop with his MAP Team.

Jeff and I eating during a brief pause in an assault on an insurgent village.

Sgt. Howard Cooke from Combat Engineers in FOB 1 in Hit.

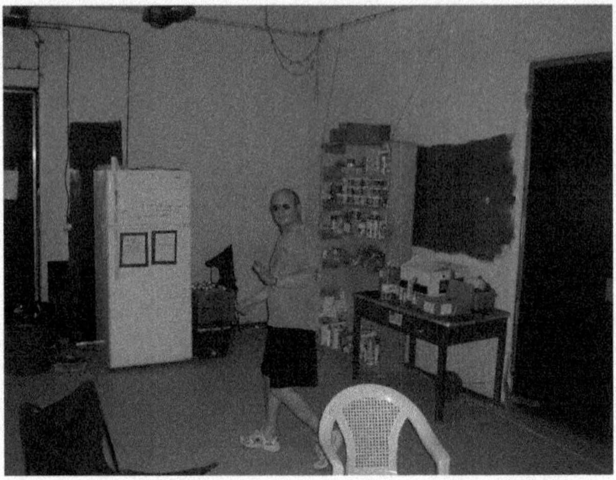

Glenn Miney inside the berthing area at the Haditha Dam.

Capt. Gratton leaving the Dam at Haditha prior to relocating to the FOB at Hit.

Matt Coleson eating rotisserie chicken inside of the FOB 1 aid station in Hit.

Troy Suza on patrol in Hit.

Thor Wold with his Iraqi Army Medic counterpart, Ganon. Ganon would serve bravely, treating many American and Iraqi wounded.

Eric Zimmerman and Reginald Bush at the Damn in Haditha.

Eric Zimmerman, Brett Schaser and Chico Nutter at the memorial service for Jeff Wiener.

CHAPTER 8
MATT

If I had to narrow down all the extraordinary moments of my deployment that had any semblance of good in them. I would give just one name: Corporal Matthew Cole, 4th AAV Battalion, USMC.

Until I wrote this, only my wife and a handful of Marines who were there knew about Corporal Cole. May 17th, 2005, his alive day, was the day of the war that I am most proud of, and until now, I kept that one just for me.

I had been again taking solace in running on top of the dam, this time in the heat of the noon sun. The days were becoming noticeably hotter, and one of the best ways to acclimate the body of a person born in the snow belt was to exercise in the heat. I couldn't hide from it, so I just needed to get used to it.

I was in front of the Chief's berthing when I felt four quick mortar rounds hit. I didn't hear the mortar rounds hit; I felt them, deep in my stomach. The concussion from the impact sent a shockwave through the air and into the ground that somehow seemed to intersect in my abdomen and combined with adrenaline a millisecond ahead of the sound of the explosion. The effect of feeling the impact served to amplify the fear that was caused by hearing the explosion. I had heard thousands of explosions in Iraq, but I had never mistaken one of them for a mortar round, and when a mortar round hit, I never mistook it for what it was, the single most frightening thing I have ever experienced. I thought about mortar attacks later, and why they were so terrible. Certainly, there were an unthinkable number of ways to die over there, but a mortar attack seemed so much more terrifying. I think it was because of the randomness and the violence that came from feeling we were sitting in relative safety.

When on patrol, or in a firefight, we had no misconception of the danger we were in, no sense of security whatsoever, and we could almost accept it for what it was. When we were back at base, we felt relief. Like we could relax and wait for the next mission, because that's mostly what the time back at our firm bases was; relaxation, boredom, and working parties to fill sandbags and clean stuff. We could almost accept that as well, that this was boring, and it sucked, but it was better than getting shot at on a patrol, and at least it was passing the time until we returned home. It was safe…as safe as we could get in a war.

I felt that safety until one second and with one explosion, I evaporated. Just like that. One minute I was missing my wife and thinking about what kind of new tuna dish I could make for lunch; the next I wasn't.

It was like God had some big mortar roulette wheel up in heaven and he gives it a spin to see who gets hit on any day. Tick, tick, tick, tick, tick, "let's see, today we have a guy eating lunch, two guys playing spades, and guy in the porta-shitter." Then boom. Mortar rounds fell from the sky, lives erased, and Marines go to heaven, or Bethesda, which was the exact opposite of heaven.

My first intimate encounter with mortars came about a month before Mother's Day. I was playing cards with some of the other Marines in my MAP team in our berthing. It was a great set up. We were housed in an actual bomb shelter built by Saddam Hussein when they built the dam. It had walls that were several feet thick, with heavy metal doors and ventilation shafts cut at right angles to deflect the blast of incoming rounds. It was very well constructed and had lasted through the Coalition bombing during two invasions. I loved that bunker. When I was inside it, I felt like nothing could hurt me. The *mujahideen* had managed to hit it several times, but only succeeded in knocking out our retrofitted air conditioning and the water to the showers. Nothing life threatening, but enough to really make me hate those guys.

We put a lot of effort into improving our position and making things a bit more comfortable, and it seemed like every time we put air conditioning in something, they blew it up. I think it was that they had lived here in this oppressive heat for thousands of years and here we come, bringing with us this marvel of technology, AC. That had to really upset them. That's one of the things that makes America the greatest country in the world, the ability to forward deploy air conditioning.

Our card game was rudely interrupted when the bunker was rocked by a single mortar round, which was accompanied by a huge secondary explosion. I didn't know what the round hit exactly, but it blew up in spectacular fashion. After a secondary explosion, I heard screaming coming from outside. Everyone in the bunker froze, trying to process what to do next. Whoomp! A second round had hit, even closer than the first.

The screams grew louder. There was a Marine trapped out there and someone needed to get to him.

Sergeant McCall, who was still in charge of K-Bar 10 at the time, yelled as he ran down the hallway, "Get away from the door, get back!" He was physically throwing Marines out of the passageway as he moved towards the entrance to the bunker to the sounds of the downed Marine. Twelve Marines, paralyzed with fear, stood in awe as Sergeant McCall seemed to move with a ferocious sense of purpose.

At that point, the entire MAP team was quite aware of McCall's Bronze Star and heroics during his first tour. The difference between his battlefield experience and our lack thereof was night and day. As soon as the rounds started hitting, we froze, and he was moving. No one had any idea what to do and Sergeant McCall's sense of purpose was reassuring. We all looked to him for what we needed to do next about the Marine down outside.

"Get the fuck out of the way! Fucking Move! Make a hole!" screamed McCall, as he ran down the passageway. We looked to him to lead us, he had been here before, he would surely inspire us to action. I expected to see a display of heroics by McCall worthy of another Bronze Star. A Silver Star even…heck, maybe even the Medal of Honor, as McCall raced fearlessly through incoming mortar fire without regard for his own safety to save a wounded Marine.

Instead, what I got was to be volunteered. "Make a hole for Doc!" Joshua McCall was tossing Marines out of the passageway not so he could run into mortar fire, but so I could.

Confused, I thought to myself, "He wants me to go out there? I'm not the war hero here, I'm not a Marine, I'm just a corpsman, why me? Why don't you go out and get your boy?"

We had just spent two months in Twentynine Palms training for war, for instances such as the one I found myself in that night. The Navy was having one hell of a time training enough corpsmen to fill all of the infantry billets with the Marines. The time and money that it took to make a sailor into an expert medical practitioner was one thing, but then to take that medical expert and turn him into a Marine was something else altogether. The Navy didn't have nearly enough Marine Corps qualified corpsmen to go around, and they certainly didn't want them getting killed doing anything stupid.

I taught my corpsmen, and any Marines that would listen, that the days of corpsmen doing a Medal of Honor run through a hail of bullets to get to a wounded Marine were over. Corpsmen were too valuable a commodity to be getting killed wholesale doing things like that. That was the Marines' job. It was a different time now.

But apparently, Sergeant Joshua McCall thought differently. After his decree that the hole being made was for me, and his quite unmistakable inference that I should have been running my ass off behind him to fill it, the twelve pairs of awestruck eyes that had previously fixed on Sergeant McCall, shifted to me like a crowd watching a tennis ball being hit back over a net. The responsibility was with me to do something.

When I arrived at the dam, I called my wife, and she made me promise that I would lay low and have no part of any heroics. I made her that promise and actually intended on keeping it. That promise, what I had taught my corpsmen at Twentynine Palms, common sense, and a good amount of fear all kept my feet planted firmly inside that bunker. "It's a different time now. Corpsmen don't go running out into fire with a 'my life be damned' attitude," I told myself.

WHOOMP! Another round had landed right on top of us. "Oh God!" cried the downed Marine. I imagined that he was feet away from where the rounds were hitting. McCall's voice was not far behind, "Doc!"

That was when it hit me. This was not a different time. This was a Marine Corps that was 229 years old, and for every one of those 229 years, corpsmen had placed their lives in harm's way to save their fellow Marines. "Through the gates of hell for a wounded Marine" was the motto. Well, this sure seemed to fit. It was decision time.

"Moving!" was all I could muster out of my cotton dry mouth. I shook my head slowly and swore under my breath as I grabbed my flack and ran out the door. Smoke and vaporized sand hung low in the air.

WHOOMP! Another round struck just as I broke the threshold of the door.

I looked to my left and saw the Marine on the ground, silhouetted by a wall of fire. The huge secondary explosion was the fuel depot being hit and going up in flames. Thick black smoke poured from the bladders that held the thousands of gallons of JP5 that fueled our battalion. The fire at the fuel depot was raging out of control and I had no idea how to go about fighting a fire in the desert.

The Marine, thankfully, was only about twenty yards from the bunker. He was down on one knee with his head tucked under his hands, screaming when I grabbed him. I didn't know how badly or where he was hurt, but I knew I wasn't fixing it right there. I took hold of the back of his flack and half drug him and half ran with him to the bunker. The whole sprint couldn't have taken more than twenty seconds.

WHOOMP!

The rounds continued to pound the area around the bunker. I dropped the Marine to the ground and started tearing off his helmet and flak jacket to start my injury assessment. I needed to figure out what wound to treat first. "Where are you hit?" I kept asking him over and over, but all he would say was, "It's so loud, God, it's so loud!" My heart went out to the terrified Marine, who couldn't have been more than twenty. I imagined that this was pretty much the worst thing this poor kid had ever gone through in his life, but I needed to find out where he was hit if I was going to help him. I grabbed his Kevlar helmet and looked him squarely in the eyes, "Where are you hit?"

Those eyes just looked back in confusion and spoke in sort of a whimper, "I'm not…I think I'm good…It was just so loud." He wasn't hit, not at all. He was screaming because he was caught out in the open during a mortar barrage and terrified. I hadn't risked my life for a wounded Marine; I'd risked it for a scared one.

I wanted to punch him, at first, I was so mad. Punch him and tell him to just duck behind something next time. Making me run out in the middle of God knows what just because he was scared. Son of a bitch!

But empathy got the better of me. I was just about that scared and I was in a bunker, heck, I might have done the same thing outside. Instead of punching him, I hugged him and told him in the calmest voice I could muster, "You're good, brother. We'll get you up to the BAS to have your ears looked at when the shooting stops… you'll be fine."

I got a round of "Way to go, Doc" from the Marines in my MAP. I shook off the compliments as nonchalantly as I could. "No worries gents, that's what I'm here for." I grabbed a warm Diet Coke and sat in the corner to take a breather and collect my thoughts. I figured if I was drinking something I probably wouldn't cry, and if I had something in my hands, no one could see them shaking. I thought about what I had just done and how I would tell Louise. I did exactly what I promised her I would not do, and it only took me a few weeks here.

Louise and I had always had a thing about promises. I had never broken one to her. Not one. If I promised, I did it. That was the rule. I felt awful. Not only for breaking my promise to her, but for almost getting myself killed doing it.

Right there in that bunker, below a burning fuel depot and with the smoke from mortar strikes still in the air, I took stock of myself and why I was here. I was here to do a job. I knew what went along with that job and what the risks were when I signed up. I vowed that I would never again stand with my feet glued to the deck while someone yelled for help. I promised myself that, above all else, I would do the right thing. That's what I was thinking when I volunteered to clear the hospital on Mother's Day. Not once after that moment was I ever standing around wondering what to do when something blew up. From that moment on, I always found myself running towards the explosion before I thought about it. Not because I wanted to be a hero; but because, like the twelve Marines in that bunker, people looked to me to do something when Marines were hurt. I appreciated Louise's request, but I figured that if I made it through this, I would have a lifetime of begging her forgiveness. And if I didn't….

CHAPTER 9
MAY 17TH

On May 17th, I passed the Chief's berthing running as fast as I could, wearing nothing but PT shorts and tennis shoes, towards the BAS which was about a hundred meters away. I arrived to find HM2 Nutter screaming at the top of his lungs. "Tracks got hit. Tracks got hit hard, right in the fucking middle." HM2 Nutter, "Chico, " we called him, was a legend. He was a Marine Sergeant during Vietnam and was wounded in the battle of Hue City. He became a paramedic afterwards and re-enlisted as a corpsman. This was his third war; he was almost sixty and had already had open heart surgery. Chico was a fine corpsman, a tremendous leader, a great friend, and the toughest man I have ever met.

Chico was relegated to permanent BAS duty, no field work for him. Not because of his age, or his ability, or anything remotely close to that, but because Chico was the oldest service member in Iraq. The whole Marine Corps knew who he was and where. The Commandant of the Marines himself even visited the dam to meet him, because the two had served together at the same time in Vietnam. He was a legend not only with 3/25, but with the entire division. Our command staff was under strict orders not to let anything happen to him. As Captain Gratton once put it, "Chico, if anything were to happen to you, a lot of heads would roll." Which was true. But more to the point was what the entire medical staff was thinking, "Plus, if anything happened to you, I could never live with myself." Which was the rest of the truth. We would never have been able to cope with losing Chico. It would have been devastating. We needed him there, like a father, watching over us. Telling us what to do, telling us that it would be okay. Keeping us safe...and sane.

Chico had a calming presence. He was perpetually writing letters. Eventually, we broke him into email, but we could always find him, when he was not working, sitting in a duck taped folding chair in front of his rack, writing and reading letters. Just as he did forty-some years ago in Vietnam, corresponding with the very same woman. It was what he did back then to get through and it was getting him through our present situation. When we saw him, we knew that he had been through this before...and if he had gotten through it, then we could too. He symbolized hope for us. He never got flustered treating casualties. The whole BAS could be screaming, and he would just stand there, look around, see where he was needed, and slide in without a word. I count him as one of the truly amazing people I have met in my life. Unfortunately, dementia took Chico shortly after we came back. He died a few years later. Chico Nutter was a good man.

At that moment, however, Chico was perched over the side of the top level of the dam, looking down to the ground, ten floors below where the tracks were housed, and where the Marines who operated them lived. Smoke poured out of the middle of the assembly of massive armored vehicles. Marines were running around and yelling, and from our vantage point, it looked like bees swarming after someone had hit their hive with a stick.

I knew from the placement of the rounds that there was no way the trackers could have escaped without injury. Marines were hurt, probably badly, and it was our responsibility to treat them.

We had pretty extensive casualty and evacuation plans for offensive operations, but back at the dam, the plan for dealing with casualties was a bit looser. After a mortar attack, the fastest guy grabbed a med bag and would run as fast as he could from the BAS to the point of impact, which could be over a half mile, and would start treating casualties. That was the plan.

Fast guy. Med bag.

That was an improvement on the previous plan which had the same essential elements except for the med bag. I added that after my last sprint that was made without a med bag. I arrived out of breath, in shorts and a t-shirt screaming for a med bag over the top of a wounded Marine. I remember the Marines who had not been hit looking at me and shaking their heads as if to say, "Great, in the middle of all of this, we get the stupid corpsman."

I managed to patch the Marine up from that first run with a bandage I stole from another Marine. I felt like a turd. I didn't have any equipment and I was wearing tennis shoes to a war. We commandeered a Humvee and drove that wounded Marine the half mile back up to the BAS and the ride took forever. I just sat there, not saying anything, looking at the injured Marine who was doing the same. So, to break the silence, I asked him, "So where ya from?" It was like turning on a faucet, he started crying so hard. "Don't ask me that!" he screamed. "That's what they ask you when you're gonna die!" He wasn't going to die, he was hit in the leg, I just felt awkward with the silence and was making small talk. Now I felt really terrible. Not only had my preparation and medical care sucked, but I also made the poor bastard cry to boot.

The minute we arrived at the BAS, I made it up to him by giving him some morphine. By the time I left, he was laughing and having his friends take our pictures together. Better living through chemistry.

Later tweaks to the running down the dam plan would be made, usually after me almost being killed or almost killing someone during my miracle mile. We added this little pearl of wisdom next: "Wear your flack and Kevlar." I learned that the hard way after yet another dash after a mortar attack in which I discovered that the *mujahideen* could fire two volleys and hit different areas of the dam. The first one hit (of course) the tracks, the second hit the dam in front of me as I was running down to them, sending shrapnel skipping off the deck next to me, and sent me diving down a stairwell. That made me rethink the whole shorts and t-shirt portion of my plan.

As I sat down to write this book, I realized that of all the stories I am telling, fully half of them occur as I am fighting a war in shorts. How about that?

So, as per the plan, the fastest corpsman at the BAS had to run down to tracks to deal with the aftermath of the most recent mortar attack. Nothing against Chico, but I figured I had the sixty-year-old heart attack victim by a step or two, so I grabbed the first med bag I found and took off. I ran down the stairwell three flights until I hit the mess deck. There I looked over my shoulder and saw Senior Chief Nadia running in stride with me. We skipped the rest of the stairs and instead ran down the next seven levels of the dam down an outside wall we called "the ramp," which was nothing more than a three-foot-wide strip of concrete on a forty-five-degree incline that let us skip seven flights of stairs at the expense of the possibility of a seventy-foot fall. A Marine in the last battalion to occupy the dam had fallen off there and almost died. On that day, however, the quickness it afforded us seemed to be worth the risk.

I arrived first and saw two Marines lying on the ground. One was yelling and holding a badly mangled arm, the other was lying face down on the ramp of an Amtrack. He was as grey as a rain cloud and not breathing. He had a hole in his back the size of my fist. That was the first time I met Matt Cole, as the life drained out of him on the ramp of that track.

I ran to him first, thinking that if he was not already dead, he soon would be. I knelt down beside him and could see that the hole in his back, made when shrapnel from the mortar tore into him, was just to one side and over a lung. Air bubbles slowly crept from the wound, telling me that if he were to have any chance of living, I needed to seal that hole with something that would force the air to go in and out of his mouth, not in and out of his back. If I didn't do that, the air going into his back would fill his chest cavity and build up pressure that would suffocate his heart and remaining lung.

I opened my med bag to grab the chest seals that we kept specifically for that type of wound. But when I looked, I found… nothing. Whoever had the bag last had not restocked it with the chest seals that I needed to fix Matt's wound.

Spectacular.

I wasn't sure that I would be able to save Matt without all of the right equipment, so I had to figure out what in the world I could replace a chest dressing with and quickly. I yelled to the Marines that were watching me work for some duct tape. Marines used duct tape to fix just about everything. I figured if it could tape together parts of a helicopter, it could probably patch a hole in a guy's back. I also knew that just about every Marine carried some, so it should be pretty easy to find. I had a roll in my hand before I even got the words out of my mouth.

I sealed Matt's wound with about five strips and rolled him over. He was still grey, still unconscious, still not breathing. Air had entered Matt's chest by way of the massive hole and was not allowing his lungs to expand properly. I had to release the air in his chest cavity in order for him to be able to breathe, and to do so, I had to put a needle through his chest wall.

Although to someone not trained to do it, a needle chest decompression can seem very daunting. I was lucky to have performed quite a few before deploying when I was working as a paramedic and was confident I could do it again. All I needed was a needle that was long enough. I reached into the med bag and found…well, not the right needles. The ones in this bag were for starting IVs and were way too short to puncture the chest muscles of a Marine who had spent the last eight years doing pushups and pullups.

"Son of a bitch, I can't catch a break," I said to myself, a little too loudly.

Too short or not, it would have to do. I opened the needle and placed it in the space between his second and third ribs and pushed. Nothing. When I inserted the needle, there should have been this rush of air and a miraculous return to life. I had nothing. So, I pushed harder, and deeper, until I was almost standing on his chest, just to be able to compress his ribs enough for the needle to reach where I needed it to go. Then it happened, the rush of air out of the needle, the rush of air in through Matt's mouth and nose, and the opening of his eyes. Hot damn! We were in business.

Matt's breathing was still dangerously shallow and his color only marginally better. He was not dead right now, but the issue was still seriously in doubt. I needed to get Matt back up to the BAS for a chest tube. The chest tube acts as a type of permanent needle, keeping air from seeping back into his chest cavity, and he needed one quickly if I wanted to keep his lung from deflating again.

There was no carrying him, or the other wounded Marine that Senior Chief Nadia was working on, up ten flights of stairs. We needed a ride. I directed some Amtrack crewmembers to grab the first Humvee they saw, and they did, but there was one problem. It was built like a Ford Explorer Sport Track, the little SUVs with a pickup truck bed just about four feet long. The bed of this Humvee was just too small to fit Matt's over six-foot frame. It was certainly not going to work.

As I pondered what to do about our transportation dilemma, Matt helped me a bit by starting to have more trouble breathing and fading back out of consciousness. The gains I had achieved with the small needle were not holding, and I would have to decompress his chest again, and move him now, if he were to live.

I stuck another needle into Matt's chest in the same brutal way I had done before. This time, Matt had the pleasure of being conscious for it. I felt terrible stabbing him in the chest, Matt just groaned in pain as I pressed all my weight onto the needle and drove it deep into his chest. With the help of two other Marines, we lowered the tail gate to the Humvee and lifted Matt inside. His legs hung over the tail gate just below his knees, but it would have to work. The two other Marines held Matt in the truck by the belt and I straddled his waist to allow me to keep decompressing his chest, which I did at least four more times during the trip.

I yelled at the driver to go, and we took off like it was the Daytona 500. In order for us to get from the ground floor of the dam up to the tenth deck where the BAS was, we had to drive outside of the base, away from the security of our snipers and machine gun emplacements and guard posts; up a mountain and then back in the gate at the top. The whole trip was done without an armed escort, at no less than fifty miles per hour, and took about five minutes.

As if holding someone half in a pickup truck at that speed, while driving up a mountain through a war zone, while I was trying to put needles in his chest to keep him alive wasn't hard enough, halfway up the mountain, the tailgate broke.

Half of it dragged behind the Humvee along with my left leg and both of Matt's legs. The other two Marines were now holding on to both Matt and I for dear life, as the tail gate bounced over rocks and slammed into us. Miraculously, those two wonderful Marines held us both.

As Matt and I were being dragged by the Humvee, I kept my hand over the needle in his chest. He reached up with his right hand, and grabbed mine, tearing it away from the catheter that was still inside him. This had to hurt him terribly, I thought, and I could understand why he wanted me to stop and why he wanted the needle out. "I know this hurts brother, but I have to keep the needle in your chest so you can breathe!" I yelled to Matt. I tried to push his hands away, but he just kept grabbing.

"Doc," one of the other Marines said to me, in a voice that was as soft and calm as if he were in church, "he just wants to hold your hand."

As I grabbed Matt's hand tightly, I panicked, and screamed, "NOOOOO!" Visions of Matt dying as soon as I held his hand flashed through my mind. I didn't want that to haunt me for the rest of my life. I didn't want that to be the war story that I told at the VFW, the time I held the hand of a dying Marine. I didn't want to have to look at his mother and tell her that even though I was supposed to have possessed this vast medical knowledge, all I did was hold her son's dying hand.

"You are not going to die! You give me five minutes!" I pleaded with Matt. "In five minutes, we'll be at the BAS. The doctor there will put a tube in your chest, and you will be fine. You just have to stay alive for five minutes. Please!" I looked into Matt's eyes. He looked so scared. Hell, I was scared for him. I felt helpless. Years of medical training and experience meant nothing. I had done everything right to save him and it meant nothing. I was reduced to yelling at Matt and begging him not to die. "Five minutes, please" I told him, over and over, tears rolling down my cheeks, as I held his hand, until we finally reached the BAS.

Captain Gratton met me at the back of the Humvee. "He needed a chest tube like five minutes ago." Captain Gratton gave me half a look like, "I'm the doctor and I decide who gets chest tubes," but that lasted only about two seconds after he saw Matt.

Captain Gratton and Chief Schaser prepped Matt for a chest tube. He could not have had two finer people operating on him. Captain Gratton's years in emergency medicine had made this old hat, and Chief Schaser's deployment with the Marines had interrupted a very fine career as a Surgical Physician's Assistant. Matt was in wonderful hands.

The chest tube went flawlessly, Matt became more lucid, and his color improved to passable. He was not even close to out of the woods, but we had bought him some time.

A thorough physical assessment of Matt found something that I had not caught previously. Matt had no mobility in his legs. The piece of shrapnel that had entered his back and landed in his lung had bounced off his spine. Matt was paralyzed.

Swarms of corpsmen reassessed Matt's motor function in his legs, just to make sure. "Movement, I've got some movement!" yelled one of the corpsmen. But he didn't, the young corpsman was only seeing what he wished he could. However, the erroneous reports spread to an audience of Marines waiting for news of their friend. "He's okay, he's moving his legs!" One Marine passed back to the next. One of the staff NCOs from tracks grabbed me and asked me if Matt was going to be okay. "He's hurt real bad," was all I could muster.

I knelt next to Matt as he laid on the helipad waiting for the medevac. "We've done all we can. You'll be in surgery in Al-Asad in twenty minutes, and they will fix you up for good. You have to stay alive for twenty minutes, it's all on you."

Sergeant Major Wagnor, the senior enlisted man for the battalion, stood next to me as the medevac lifted off. His face was as white as mine. "He gonna make it Doc?"

"I don't know; I've seen guys hurt worse live…but I've also seen a lot of guys hurt that bad die. I guess it's not up to us now, Sergeant Major," I replied.

He walked away with his head down. The man could be a son of a bitch, but every time someone got killed or wounded, he took it like it was his own son. He would earn a hand full of Purple Hearts himself during that deployment. But I think if he had the choice, he would have taken getting wounded about a thousand times rather than see one of his boys get torn up like that.

Matt lifted off for Al-Asad, alive. Not in the best of shape, but he was alive. "Nice job saving his life," Captain Gratton said to me in a voice that was obviously trying to conceal the emotion and exhaustion of the moment.

I'd been in the business of saving lives since I was eighteen and I could say with certainty that although I had been involved in saving many lives, there never had been one that I could claim sole responsibility for. Never someone that I could say, "Yeah, you see him…I saved his life…No big thing, it's just what I do." I walked around for the next day and a half, as proud as a new father, with what I had done. I saved a life damn it. Me, all by myself. Hooray for me.

That was it. I was convinced that's why God let me survive Mother's Day, and maybe the entire reason for sending me to Iraq was to save Matt Cole's life.

Humility can be a hard learned lesson.

Two days after the attack, in our morning briefing, HM1 Dibell, our battalion's casualty tracking officer, gave us all some grim news about Matt. He had taken a turn for the worse in Al-Asad. A doctor who was assigned to a Navy construction battalion, and also a reservist, had happened to be in Al-Asad when Matt arrived. Thankfully for Matt, this doctor was a Cardiothoracic surgeon for his day job and was more than happy to open his chest and stop him from bleeding to death. From there, Matt had been sent to Germany where it was predicted he wouldn't survive. We were told his family had been flown there, awaiting the end. Saving someone's life doesn't count if they die anyway. I had little to do now with whether Matt lived or died. It was up to God, Matt, and some doctors that had never met any of us.

I was crushed. I thought I had perspective on my reason for being here, I thought I had done some good. I thought I had lived to fight another day, but instead I had just lived to have another part of me die. I probably took the news about Matt as hard as Mother's Day. We all did. Marines had been dying and finally we saved one… for a little bit.

I was immediately ashamed of being such an egotistical ass, boasting about what I had done. If I was walking around taking credit for saving his life, shouldn't I be the one to blame if he dies?

It would be days before we got word on Matt. He would live but would be paralyzed. I remember that this news didn't comfort me as much as I had hoped it would. I thought that hearing Matt would live would lift this giant weight of hopelessness off of my chest, but it didn't. It helped ease our grief some but ended up giving us a more pragmatic view of our roles in this war. We would certainly do some good. But even on days where we did our best, the end result could still be a young Marine chained to a wheelchair for life. It was hard to count that as a win.

I hadn't saved Matt's life. I helped, but Captain Gratton and Chief Schaser saved his life with that chest tube. And if I asked them, they would deny it. They would say that the thoracic surgeon who happened to be at Al-Asad saved him. And if I asked that thoracic surgeon, he would say it was Matt's iron will and commitment to life that saved him, a very valid point indeed. And if I asked Matt, I'm sure he would say that he had little to do with living or dying, that he would say that decision was up to God.

A few years later, I got the chance to ask Matt that very question. I had been trying to track him down from the very minute I got home. I talked to Marines who had been in his unit, made phone calls and sent countless emails. No luck.

Thank God for Google. A couple years and a thousand searches later, I found Matt's address and sent him a letter. With his permission, here is part of it:

"Matt,

I am sure you have no idea who I am. We have only ever spent about 45 minutes of our lives together. I will tell you though, that I have revisited those 45 minutes every single day since May 17th, 2005.

My name is Taylor Cleveland. I was a corpsman for 3/25 who served with you at the Dam in Haditha. I was the person who took care of you after you were wounded. It has taken me almost two years to find a way to contact you...and another two months to fight through the tears to write these first two paragraphs. I hope that by writing you I am not opening wounds that you have been fighting to heal. I just wanted to thank you for what you did for me that day. Because you see, the day you were wounded, you saved my life...That was the first time we met. This is what I remember about that day..."

I went on to retell the events of that day, believing that he should know how truly miraculous the circumstances that led to his survival were.

"I really feel that God never gives us any more than we can handle. He knew what I could handle in Iraq, and he gave it to me.

I told you earlier that you saved my life, and that's just what I meant. If you had died holding my hand in the middle of that desert, my life would have been over. I never would have emotionally or psychologically been able to recover from that. That would have been more than I could handle. I would have been another casualty of the war. I may have made it home, but the life, the home and the family that I have been blessed with would have been cast off as memories of the person I was before the war. You saved me from that. You lived. Because of your strength, your determination and intrepidity, you lived. And doing so, you allowed me to keep living.

Thank you Matt. Thank you for the life I have today, and all you have given up so that I could have it.

If there is ever anything that I can do for you, or your family, I would be honored.

Semper Fi,
Taylor "Doc" Cleveland"

I got a call from Matt two days after I sent him the letter, the very day he received it. Matt had been paralyzed from the waist down.

Coincidentally, on the day he received my letter, he also received a letter informing him that he had been chosen for an experimental surgery that had proved very promising in restoring the ability to walk in paraplegics. Matt got both letters, however, after returning home from a memorial service for his best friend who had been killed when his Amtrack hit a mine while we were in Iraq. Life is full of some good moments and some bad moments, but mostly, it's full of moments that fit into both categories at once.

Matt and I talked for about an hour. I managed not to cry, which was amazing. He thanked me, and I thanked him. We caught up on our lives and our families. There were times when neither of us could think of anything to say and there would just be silence. But the silence was never awkward. It was comforting. It was like we were just thankful to be able to sit in silence together and reflect. As we sat in silence, I imagined the emotion on his face. It was the same that was on mine.

After a particularly long stretch of silence, I managed to ask him a question that had plagued me.

I was thinking back to that bus ride before we deployed, when the Marine asked me earnestly to not let him live a life in a wheelchair, saying that he would rather be dead than live like that.

Now, I wondered if Matt had made that same request with his corpsman on some bus ride before he deployed.

So, I asked him, "I wonder, a lot, if I did a good thing. I wonder if there is ever a time that you wish you didn't live."

The quarter second that it took for him to speak passed like a year. His answer was short and sincere.

"You did a good thing."

"Thanks, Matt. That means a lot."

I lied. It didn't mean a lot. It meant the world. It meant a thousand worlds. Those five words were probably solely responsible for my emotional health today. Thank you again, Matt.

Matt's wife, Kim, and his father-in-law came to visit me that summer. We talked. Some about the war, some about that day, but mostly about what comes after. We talked about mental health and the distinct lack of it for almost all the Marines from 3/25 who returned.

I told Kim that I just came home a few days earlier from Justin Mangano's funeral. Kim seemed sad, but not a bit surprised. Two Marines from Amtracks had killed themselves, and she had a list of others who could easily follow suit.

I talked with her about some ways that I have managed to cope with things, people I have talked to, things that I do to stay right with myself and my family. It was very comforting for Kim to tell me that Matt was doing many of the same things, and that he was now doing well emotionally.

Hearing that one of my friends is having emotional problems after the war is, and I feel terrible for saying this, bittersweet. One part of me is devastated that someone so close to me is hurting. But the other part of me is relieved to know that he felt the same way I did…that I am not alone.

Kim thanked me. She actually said the words, and I could tell they were hard for her to get out. She didn't have to say them; her visit was more thanks that I could ever ask for.

Today, Matt and Kim have a crew of the most beautiful children imaginable. Kim is a teacher and Matt is working with a group that helps disabled veterans start farming. Theirs is the first Christmas card we get every year. Matt Cole is a good man.

CHAPTER 10
MINEFIELD RUN

My first mission after being cleared for duty after Mother's Day was in support of Lima and Kilo Companies, as they retook the city of Haditha. After Mother's Day, it became readily apparent that there was a whole mess of insurgents in the city, and the plan was to go door to door and root them out. Since the last time the Marine Corps went into that city and we got our nose bloodied, the assumption was that we would be taking rather heavy casualties. To prepare for this eventuality, I and two other corpsmen would be placed closer to the front line in an arraignment called a Forward Aid Station (FAS).

The FAS concept had started during the battle of Fallujah. The Battalion Surgeon for one of the units involved in the assault, in creating a casualty evacuation plan, realized that the evacuation time during urban fighting could be significant. To combat this, he took a small contingent of medical providers and placed them as far forward as the battle tempo would support. Doing so would reduce the time from injury to definitive medical care and would, hopefully, reduce the number of fatalities. The FAS concept was implemented and worked fabulously. Medical providers were able to provide advanced lifesaving care almost immediately after injury.

Prior to our deployment, 3/25 studied the lessons learned from previous units who served in Iraq, including the battalions who fought in the battle of Fallujah. One of the many things we took away from this study was the FAS concept, which we had employed regularly with success, and were planning to do again in Haditha.

I was happy to be doing something useful again, and the truth be told, FAS duty had not proven to be all that dangerous for the corpsmen of 3/25 so far. Granted, it was not as safe as being back at the dam, but I wasn't kicking in doors with a line company either. I thought that the mission was going to be good, something to get me back in the game, gently.

The thing was, there was nothing gentle about shooting people and getting shot at. It was life or death, and it was always serious. The second I thought otherwise, I was proven wrong. The assault on Haditha was no exception to this rule.

The command element, to which the aid station was attached, inserted into the outskirts of the city and established an operations post at an abandoned school. While we were doing this, assault elements entered the city from all directions. The assault elements took fire immediately, and within the first moments of the battle, two Marines, Sergeant David Winberg from Lima Company, and our FAC, the officer responsible for directing aircraft flying overhead to targets on the ground, were critically wounded. What was supposed to be a two-day operation was shaping up to be an awful engagement.

My good friend, HM2 Vic Lewis, was the corpsman assigned to Lima's assault element, and was responsible for caring for Sergeant Winberg and the FAC. When the FAC was hit, he was shot in the pelvis and dropped in the middle of the street. Vic had to return fire and fight through a hail of bullets to reach the injured officer. Once he got to him, Vic pulled the man to a position of cover and started treating him, all while still being shot at. Vic then carried the FAC back to the command element, where he continued to treat him and Sergeant Winberg, until medical evacuation arrived. For his actions, Vic was awarded the Bronze Star.

The care that the FAC and Sergeant Winberg received from Vic was remarkable. The problems started with the medevac. The plan called for all casualties to be ground evacuated first to the FAS, and then either evacuated by ground to the BAS for further care if they were not life threatening or evacuated by air to Al-Asad if they were life threatening. It was a good plan. But like all good plans, if they are not communicated well, or not understood or followed, then they don't work. And in this case, the plan didn't work.

When Sergeant Winberg and the FAC were wounded, a decision was made by Lima Company to request air evacuation directly to their position to extricate the wounded Marines, instead of transporting them to the FAS. I don't know why that decision was made. Maybe there were no ground assets available for transport, maybe they didn't know the plan to transport all of the casualties to the FAS, maybe they didn't understand the importance of transporting all of the wounded to the FAS, and that air assets would not be available to fly directly to them. Whatever the case, the Marines laid bleeding in the desert for forty minutes, begging for medical evacuation helicopters to land, while being told the fighting was too heavy.

During the planning stages, we knew this could be the case, and used the FAS to mitigate that risk. Eventually, the fighting subsided, and a medical evacuation helicopter was able to land and transport the two wounded Marines, but it was too late for Sergeant Winberg. He had gone too long without surgical care. He died in that city, waiting for help.

I was part of the planning for the medical component of the assault. As part of that planning, I should have made sure that the combatant commanders understood what we were asking of them and why. We also should have made sure that they would agree to follow the plan and not go off script, and that they understood the consequences if they did. If we had done this, Sergeant Winberg would be alive. Sergeant Winberg joined the Marines to serve his country and to help people. When his deployment was finished, he wanted to become a firefighter so he could continue helping people and serving his country. Dave Winberg was a good man.

The school that the headquarters element had set up shop in

provided a good staging area for an operations center, with plenty of room to treat patients. It also had an added bonus of a courtyard that was walled in with enough room to land a helicopter directly outside the building, but secured inside the walled area, to evacuate the seriously wounded with limited risk to the aircraft.

Immediately after we established our aid station, we started treating casualties. Thankfully, most of the casualties we were treating were wounded insurgents, with an occasional minor wounded Marine. One of the first insurgents I treated was a man about twenty-five, with a gunshot wound to his chest. The round had entered below his right nipple and had exited out of his back. I did a thorough assessment of the man and found that despite a very severe wound, he wasn't doing all that bad. I dressed the bullet wounds and spoke with one of the Marines from Lima Company who had brought in the insurgent.

"What did this guy do?" I asked. I half expected the Marine to shrug his shoulders, meaning that the Marine didn't know the circumstances surrounding the Iraqi's injury, just that he had been wounded. That had become sort of the standard answer whenever I asked.

I still asked because I was trying to determine if the patient was a good guy or a bad guy. It was so God awful hard to tell them apart. The Marines only shot people they thought were insurgents, but at the end of it all, it's not like they carried a card that positively identified them as a bad guy. If we didn't find them with a machine gun or explosives on them, then who the heck could tell for sure.

The answer the Lima Marine gave sort of surprised me: "He tried to shoot us with an RPG." Generally, guys who did that didn't make it to the aid station.

The Marine walked slowly back to his unit, as I sat there staring at the wounded man. Had a Marine received the same wound, I would have landed a helicopter in the courtyard ten minutes ago, and he would already be at Al-Asad being prepped for surgery. But given the circumstances, I had no sense of urgency in evacuating the insurgent. I bandaged his chest in a way to relieve the pressure of air building up in his chest cavity and placed him in a position where he could breathe more comfortably.

I have to admit at the time, treating the insurgent didn't seem like as much of a moral dilemma as it does now. That day I remember thinking to myself that the man was human, and I had an obligation to take care of him. I also remember thinking to myself that this guy could have been shooting at me on Mother's Day, or worse, he could have killed Jeff. I kept those thoughts in the back of my mind as I slowly made my treatment decisions. I might have an obligation to treat him, but I don't have an obligation to bust my ass doing it.

I gave the report of the wounded insurgents to the Operations Chief at the COC for medevac decisions. "They can all go by truck when K-Bar 10 gets here," I told the watch officer, assuring him that no helicopter would be needed.

"What about the guy shot in the chest?" the Ops Chief asked, double checking my assessment.

"The guy that tried to shoot Lima with an RPG? He doesn't look like he is trying to die. His vitals are good. He should make it for an hour or so." It was an honest assessment. Besides, to call in a helicopter during an attack would be a risk to the flight crew. There also weren't that many helicopters to go around and if we tied up one of our air assets when it wasn't really needed, that meant a Marine who did need it could die.It was funny, that day. I could really have given a shit if that guy lived or died. If he took his last breath as I was eating a package of Lornadoons, it just didn't matter to me. I had patched him up, sent him out, and the rest was on Allah. Until I wrote this, the thought that I should have done more had never crossed my mind. I didn't feel then that I needed to justify not flying him out, I just didn't care that much.

I thought then, as I do now, that always doing the right thing and having the moral high ground was important. I guess as I write this it was obvious that my morality sunk to a record low that day.

The pace remained steady, but not intense. Around noon, a lull in the action gave us a chance to grab lunch, restock our supplies and reassess our operations. The other two corpsmen and I were talking just inside the door of the aid station when a massive explosion filled the room with sand and knocked us to the ground.

Whatever had blown up, I thought, blew up directly outside of the door to the operations base, smack in the middle of where the supply convoys were staged. I looked out the door and could see a cloud of dust and smoke. I scanned the courtyard for any wounded when I heard frantic screams from the other side of the wall. "Corpsman up! Corpsman up!"

I grabbed my flak jacket and sprinted out in the direction of the screaming. On the outside of the courtyard wall, about a hundred meters away from the aid station, an Amtrack had hit a landmine. There was smoke pouring out of the crippled vehicle and one of the crew members was standing on top yelling for a corpsman. I ran along the inside of the wall a hundred yards to the blast site, figuring that when I got to the Amtrak, I would jump the barrier wall and be able to get to work.

Now sprinting a hundred yards wearing seventy pounds of gear in over a hundred-degree heat may sound hard, but it was ten times as bad as it sounds, and when I got to the wall on the other side of the blown up track, I was spent. To make matters worse, the wall I had to scale was eight feet tall, and I'm only six feet tall. As hard as I tried, there was no way I could make it over. I would have to sprint back another hundred yards to the gate, exit the compound and run back that same hundred yards yet again, to get to the wounded Marines. By the time I exited the gate, I could hardly run any more. My lungs were burning, and I thought I was going to throw up.

I started to slow to almost a trot when I saw the track crew member standing atop his smoking vehicle, waiving and pleading for someone to help him. "Corpsman up! Move your ass, Doc!" Adrenaline took over and my heart felt like it was going to explode as I legged out the final feet to the wrecked armored vehicle. By this time, several other Marines had emerged from the compound and were headed out the gate to offer what help they could.

As I reached the Amtrak, the Marine on top pointed at the crew of Marines exiting the gate and yelled to me "Tell them to get back, we're in the middle of a minefield!" I turned and passed the word sharply to the other Marines "Stay there, this is a fucking minefield!"

The second the word "minefield" crossed my lips, I realized what I had said. I turned quickly to the Marine who had given me that little pearl of wisdom and let him know that I was anything but pleased. "You had me run through a fucking minefield? Are you fucking nuts?" He just shrugged his shoulders and got sort of a "what are you gonna do" look on his face. A minefield. At least he could have apologized.

The Amtrack was a wreck, both tracks were blown off and it didn't look like there was one piece of equipment that was operational. By some miracle, the two Marines inside were not dead. The blast had broken the legs of a very lucky lance corporal and given a rather serious concussion to the gunnery sergeant in charge. After I treated both of the wounded Marines inside of the vehicle, my focus turned on how to get them out. Carrying them out one at a time in a fireman's carry through the minefield didn't seem like an option.

When I climbed out of the hatch and on top of the track, I saw the four Marines whom I had warned to stay out of the minefield. They hadn't listened. I ran through a minefield to help wounded Marines because I didn't know it was a minefield. They did it, knowing it was a minefield, just to help their friends. That's loyalty.

We loaded the two wounded Marines onto litters and very carefully retraced our steps back through the sand. That was how the four Marines had safely reached the track, placing their feet in my footprints; figuring that if my step had not triggered an explosion, then there was a good chance that spot was safe. We would do the same on the way back.

So, at the end of the whole event, the score was one Amtrak blown up; the end of the war for two very alive Marines, and the living crap scared out of one minefield sprinting corpsman. That was about as good a day as I could ask for, all things considered.

The one thing I can say for the insurgents was that if we stayed in one place for any period of time, they were bound to attack us in some form. This proved to be true for the Marines in Haditha, as well.

We had just finished evacuating the injuries from the mine strike and were taking a minute to regroup when the radio cracked to life with a call for assistance. Kilo Company, who served as a flanking unit on the outskirts of the town, had come under attack while they were staged in a building. Insurgents had surrounded the building and peppered it with RPG and small arms fire. One Marine had been killed with several more wounded. The encircled Marines called for immediate reinforcements and a medevac.

Our COC had received the call for assistance and had assembled a squad of Marines from Lima Company in an Amtrak and a MAP team, to reinforce Kilo Company and evacuate the wounded. As the Lima Marines loaded onto the Amtrak, I wanted to speak with their corpsman to review the casualty evacuation procedures and make sure that he had all of the equipment he needed.

When I looked for the corpsman, much to my surprise, I was told by the squad leader that they did not have one available, but that they would be more than happy to have me ride along.

The Kilo Company position was only blocks away from the FAS, and when I walked into the courtyard, I could see the battle raging. Tracers poured into the insurgent positions surrounding the building. Tanks leveled houses with main gun rounds, and Marines screamed into the radio directing fire into the attacking forces. As I looked out over the city, it looked like the first scenes of a very bloody war movie.

I wish I could say that as I grabbed my bag and hopped into that track, I had moved past being afraid. But the truth is, I was very afraid. The last time I rode in to save the day for some besieged Marines, a whole lot of people, including my best friend, died; and I got shot in the grape. I didn't want to go through that again.

The Amtrak started taking fire immediately after we left the COC. The armor plating on the track afforded the dozen or so Marines inside protection from "most" small arms fire. One of the big problems with Amtracks though, was that the heavy weapons mounted on the turret only protect one area at a time, making the track vulnerable to attack from the sides and back. How did the manufacturer solve this problem? Well, they cut two big holes in the top of the thing and put hinged lids on them, sort of like a moon roof. Then, four of the dozen Marines who just jumped inside to be protected from the flying bullets, open the moon roof, stand up, and stick the upper half of their bodies outside the vehicle, pointing their rifles in all directions to try and "protect the track". I still marvel at the idea that flesh and blood Marines ended up protecting the vehicle that was meant to protect them. Riding in an Amtrak was certainly quicker than walking, but it was only slightly safer.

That day, I sat as far from the moon roof as I could get, listened to the rounds slam into the side of it, and quietly said a prayer. I asked God to take care of me, and to give me the strength to do my job...come what may.

The Marines posting security on the top of the track and the MAP team that accompanied us were engaging targets at will as we screamed towards the Kilo Marines. The MK-19 and M-2, .50 caliber on the turret of the track were firing controlled bursts in tandem and approaching buildings. After each burst, an entire floor of a building would disappear in a cloud of smoke and fire. Sometimes that inferno evaporated insurgents in their firing positions, sometimes insurgents walked right through the clouds of smoke and dust and continued firing, like the rounds from our heavy guns had passed right through them with no effect.

As we approached Kilo's position, the vehicle commander briefed me on the plan of action. "Doc! We're going to pull in front of the building and drop the ramp. You just go do your thing, get those Marines and we'll be here for your ride out." That sounded like an excellent plan. Every Marine with a gun was shooting at something, and I figured the more of them that were shooting, the less bad guys there would be to shoot at me when I was carrying out the wounded.

Small arms fire continued to bounce off of the armor plating as the vehicle commander yelled towards me "Ready, Doc?" I nodded in agreement, and the ramp started its descent. As it dropped, I could see that there was about fifty yards between the track and the building that I would have to run while under fire.

That meant for every wounded Marine, round trip, I would have to cover 100 yards under fire. There were at least four wounded, so that meant running ¼ of a mile while getting shot at. Spectacular.

Once the ramp started the drop, and we saw how far away from the building we were, everyone in the track started to do the same mental math that I had done. And everyone came to the same conclusion that this was probably not going to end well for me. Seeing this, one of the privates from Lima, a kid who looked to be about five-foot-three and a hundred-thirty-five-pounds and fourteen years old, strapped up his helmet, grabbed his rifle and announced eagerly, "I'm going with Doc!"

To put this in perspective, about fifteen minutes ago I was almost too scared to do the job that I had signed up to do. This kid had only spent about three days as a civilian after he graduated high school. After graduation he went right to boot camp. From boot camp directly to the School of Infantry. And from the School of Infantry directly to Iraq as a combat replacement for the Marines were losing. Months ago, he was a child. He was in history class, and his room was a mess, and his mom poured his cereal for him, and he never folded his clothes and he had to be told five times to take the garbage out. He was a child just a few months ago.

But today, he was volunteering to run through a hail of bullets with me. This wasn't his job. I didn't know this kid. He wasn't my friend. He was so new that he hadn't made friends with anyone yet, let alone the Marines from Kilo Company who he might never meet. He volunteered because I needed a hand, and some guys in that building needed a hand. And because all of those guys wore an Eagle Globe and Anchor on their blouse. That was all the reason he needed to risk his life. That is the definition of sacrifice. I never knew his name, but in the months that followed, I would see that same spirit about a hundred more times in Marines that had been children only months before. Those Marines were good men. They were the best of men.

After the ramp dropped, the young private and I both took a

deep breath and sprinted towards the front door of the three story building. I slid through the doorway like I was sliding into home plate, and came to rest next to a Marine who was only inches away from me engaging targets with his M-16 from a first floor windowsill.

I popped up to one knee and put my rifle into my shoulder and aimed out of the same window. I could see many muzzle flashes coming from a two story building about two hundred meters away. That clearly was where all the incoming fire was originating from, and I started firing in concert with the other Marines who were trying to stop it. M-16 rounds, SAW rounds and .50 caliber rounds from the Amtrak raked the building, but could not seem to stop the fire from the insurgents inside.

I paused to reload with a fresh magazine when I heard an officer yelling "Cease fire, cease fucking fire! We're gonna drop a bomb on that thing!" Excellent idea. A thousand-pound bomb from an F-18 usually stopped any fight. Although I didn't quite understand why we had to stop shooting in order to make that happen.

I looked over to the Marine next to me and passed on the order to cease fire and was shocked with who I saw.

"Cleveland?" asked the Marine, who was actually not a Marine, but rather a corpsman.

"Ocampo?" It was Alvin Ocampo. Ocampo was a surgical technician for a plastic surgeon in Texas. He made his living in the civilian world assisting doctors with breast implants, tummy tucks, and penis elongations. He was a reservist who was activated during the initial invasion and had volunteered for a second tour with 3/25. He was a great guy, easy going and just fun to be around. We hadn't seen each other since Twentynine Palms, and amidst all the shooting, we took a minute to catch up. Some of the places you run into people you know…wow. During the insurgent attack on the building, Ocampo had suffered a significant traumatic brain injury from an RPG explosion. In spite of that, he stayed in the fight. Ocampo would continue to stay in the fight for the next couple of months before several more explosions caused a series of successive concussions that made it impossible for him to continue fighting. Alvin Ocampo is a good man.

Our reunion ended when I asked where the casualties were. "Follow me." We walked up a set of steps to a hallway that had a gaping hole in the wall, obviously caused by an RPG. Lining the hallway were several injured Marines and the body of Major Ricardo Crocker. Perched atop the Major's body was HM2 Kendrick. Kendrick was a quick, smart corpsman who had drilled with Kilo Company in Moundsville, WV. These were his friends. As he stood over the dead and injured, he reminded me of a grizzly bear protecting her cubs.

"What's the scoop?" I asked, kneeling down next to the Major. Kendrick gave me an unpleasant look, like I had three seconds to leave him, or he would shoot me where I stood. With as much sarcasm as he could muster, Kendrick replied coldly, "He's dead."

I felt terrible for Kendrick, and for his Marines. They lost a friend, and he wasn't coming back. In a minute, I would take the major's body away, and that's the last time any of them would see their friend. As much as I was there to help, when I took the major, it wasn't going to be a pleasant memory for them.

I put my hand on Kendrick's shoulder, "Kendrick, it's Cleveland. I'm your medevac." The look on his face turned from rage to relief as he calmly gave me the casualty report. "RPG hit the building. The major is dead. I got two urgent surgicals with shrapnel wounds and Sabo has a head injury." Jim Sabo was another of the Kilo Company corpsman, and a paramedic in Pittsburgh for his day job. I did a quick assessment of the wounded. I ran through a checklist in my head as I looked over the Marines. Shrapnel wound to the back, needed to go in a hurry, mild shrapnel wounds, he could walk, the major would have to be carried. Sabo…

Sabo was sitting in the hallway motionless and speechless. I asked him how he was doing and where he hurt. He looked up at me and tried to form a sentence but couldn't. Jim looked like he had his bell rung but good.

The lance corporal that came in with me helped me carry the major into the track first. Next, we went back for the Marine with the shrapnel wound to the back and laid him on the bench seat. We ran back to get the final two patients only to see Sabo and the other Marine helping each other up the ramp.

A few short seconds later, the ramp slammed shut and we were off. I called Bulldog, our COC, and gave them a casualty report.

"Bulldog, Bulldog. Echo-Five-Charlie. CASREP."

"This is Bulldog, send it Echo-Five-Charlie."

"I have one KIA Marine, two urgent surgicals and one priority, enroute to the FAS request immediate helo evac. Over."

"Bulldog solid copy on all, your air is enroute. Out."

The casualty report allowed for the COC to have the helicopter at the FAS shortly after we arrived, giving us only minutes to provide care at the FAS once we arrived.

I went to work on the most seriously injured Marine. The track was flying down the road to the FAS, hitting every pothole along the way and Marines from the Amtrak were still firing into the as-of-yet not blown up house trying to suppress the insurgents who were still shooting at us. Amidst all the fire and commotion, I tried to visualize and treat the shrapnel wound in the Marine's back, but I could barely stand up, let alone find the pin hole that threatened his life.

As we pulled into the FAS, I was still looking. I yelled to Chief Steele, the chief corpsman at the FAS, that I needed help. We laid the Marine on the ground and started a more thorough examination. We weren't thirty seconds into it when I heard a report from the Marine who was controlling the landing of the medevac helicopter. "Doc, Anchorman is on deck!"

Anchorman was the call sign for the Army Blackhawk stationed in Al-Asad who was responsible for the bulk of the casualty evacuations for 3/25. I never met the crew, other than the brief interactions I had with the flight medic. I wish I could have. I would have loved to thank them. They were tremendous.

The helicopter had landed in the courtyard under precarious circumstances. There was still a heck of a lot of shooting outside. The second the aircraft hit the sand we were loading the wounded inside. Treated or not, the helicopter had only moments on the ground, or it risked being shot to pieces.

Sabo climbed aboard last. I yelled good luck to him as he lifted off, but the look on his face told me that he didn't get the message. Between the emotion of the day, and the fact that his brain had been knocked around the inside of his skull like a hockey puck, you could tell the lights were on, but no one was home.

I looked around at the FAS, now empty of patients. Snipers and Lima Marines on top of the COC were still engaging targets only yards outside of the COC. In the distance, F-18s dropped bombs on insurgent positions. The night sky lit up in some rather spectacular orange hues.

The battle continued on. I sat down, took a drink from my camel back and thought about the men who I had just flown out.

Major Ricardo Crocker was a reservist, and a Police Officer in Santa Monica, California. When he wasn't serving his city as an officer, he volunteered his time coaching Police Athletic League teams for kids in the toughest of neighborhoods. In Iraq, Major Crocker was a CAG or Civil Affairs Group, Marine. His job was to offer reconstruction help to the cities after we had driven the insurgents out. Even in another country, four-thousand-miles away from home, Major Crocker was intent on making neighborhoods better for kids to grow up in. Ricardo Crocker was a good man.

Jim Sabo was an outstanding paramedic for the city of Pittsburgh EMS. Tough, funny, and smart, he made serving the citizens of his community his top priority. In 2007, Jim was awarded one of his department's highest honors for saving the life of a victim of a house fire. Unfortunately, Jim continued to struggle with the wounds he received that day. As many of us would learn, traumatic brain injuries, big or small, would be some of the longest lasting, and most devastating injuries. Jim Sabo is a good man.

I had been running, shooting, fighting and treating wounded for almost an entire day. I was done. As the rate of fire from the rooftop started to slacken, I closed my eyes and went to sleep. I don't know how long the fighting lasted that night, but when I awoke, the COC and the city surrounding it was a picture of serenity.

People were eating, reading, walking around with no shirts on. No one was shooting. I wondered just how long I had been asleep. I checked in at the COC to see how the operation was going, and it appeared things were winding down.

After about a week at the FAS, we pulled up camp and went back to the dam. We had run the insurgents out of Haditha, and like we predicted, took our lumps doing it. Now it was time to leave and go clear them out of the next town. As we rode out, I wondered how many times we would have to ride into Haditha, shoot things up and leave before someone realized that this was not the best way to win the war.

CHAPTER 11
HIT 1

For the last three months of my tour, I was assigned as the chief line company corpsman for India Company. As the tour wore on, what was apparent to everyone in the entire battalion who had been getting shot at, blown up, and generally kicked around for the better part of four months, became a realization by those in charge: we didn't have enough guys to do our job.

In actuality, we needed about another couple of thousand Marines to effectively control our area of operations. Even then it would have taken a while to retake the entire theatre, because we had not been able to hold any of the ground we seized to date.

So, in typical Marine Corps fashion, the commanders of the 2nd Marine division responded to our need for thousands of additional Marines with about a hundred and seventy-five guys. India Company, my boys from Buffalo who had been posted as base security for the air base at Al-Asad, were being relieved by a National Guard battalion and were finally going to get into the fight.

I felt bad for the India Marines. By a crappy stroke of luck, infantrymen who had trained for months to find terrorists and kill them, were relegated to guard duty. Not that guard duty was not important. Al-Asad held some of the most important facilities in our theatre, namely a Burger King, a Pizza Hut, the dining facility (famous for lobster Mondays), the gym (complete with personal trainer) and the Morale and Welfare Center that hosted salsa dancing nights on Thursdays.

Combat was a foreign concept to that place. When I rolled in on a medevac with my MAP team, K-Bar 10, we would stop at the PX after dropping the wounded Marine off at the hospital, to load up on cold Gatorades. After one of those shopping trips, we came outside to see a group of five squeaky clean Marines with fresh haircuts sitting on top of our shot up, blood-stained Humvees, taking pictures of each other.

Yes, all of this critical infrastructure was so important to the support staff that lived in air conditioned trailers, that they thought it imperative that some of the best infantrymen in the entire country should be held back to guard it from being overrun.

There were a lot of things that made me scratch my head about the way the war was conducted, but in fairness, I haven't been to a military academy, or a war college, and I just don't have a lot of big picture perspective to second guess decisions that were made. But the one thing that was absolutely criminal, was keeping a fifth of the most engaged infantry battalion in the country on the sidelines at Al-Asad when the rest of its men were screaming for help and getting the crap kicked out of them. That was just wrong. And even though I can't attribute a specific incident to it, I believe in my heart that doing so costs lives.

Kilo Company was operating out of a FOB near the city of Hit (pronounced heet). They were several kilometers outside of the city, and tasked with the job of keeping major roads and intersections open for re-supply. For the first half of the tour, this job was a fistfight for the Kilo Marines. The city of Hit, a town of about sixty-thousand people that had been largely ruled by insurgents and untouched by American forces. Some Marines in the battalion that we replaced had conducted a patrol through the city on one occasion and were summarily thumped by what I'm sure seemed like all of Al-Qaeda in Iraq. After hearing that, Kilo didn't venture in there.

There was no sense picking a fight when you knew you didn't have enough guys to win it.

As such, the insurgents used the city as a base to launch near constant attacks on the Hit Marines. A sign was posted in the BAS on a dry erase board "Days without a mortar attack," followed by the number of days since they had last been shelled. The number never made it out of single digits. It was so disparaging to see that sign in fact, that Marines instead started counting the number of incoming rounds received to date. That number, last I saw, was in the hundreds.

When I was at the dam, at least once or twice every week we would get a briefing about another attack on FOB Hit that resulted in multiple Marines being evacuated. I thought to myself at one point that all of Kilo Company must have a Purple Heart by now.

Well, apparently, someone of some rank had just about enough of letting the *mujahideen* have an entire city, so they decided to take Hit back. To accomplish this, Lima Company, Kilo Company and some Army guys from the 1st Armored Division and a company of Iraqi Army soldiers were sent in to clear the city and search every house. This was billed as the biggest operation our battalion had undertaken to date. We were supposed to take it over, keep it safe, rebuild it and turn it over to Iraqi control. Sort of a lot to accomplish in less than a pro football season.

Once the town had been cleared, Kilo Company would take over a building on the south side of the city and conduct security patrols in half the city to ensure that the insurgents didn't return. Lima Company would do the same on the north side until they could be relieved by India, who was at Al-Asad getting spun up for their mission.

At the dam, the battalion was planning for the mission, and detaching a significant portion of its assets down south to help with the assault. Captain Gratton assembled all of the medical section and briefed us on our operational responsibilities. First, the battalion medical staff would set up a forward aid station to treat casualties from the assault at the point of attack. Corpsmen from the BAS at the dam and FOB Hit would be used for this part of the mission.

The second part of our mission was to staff India Company with corpsmen to support their combat operations. I was to be the Chief Company Corpsman, responsible for all the medical aspects of India Company as it operated within the city. With the type of operations our companies were conducting, we needed at least nine corpsmen to ensure that every unit maneuvering in combat had someone to patch them up when they got shot. India currently had two, Coleson and Goekler. I made three. I would get two more corpsmen from FOB Hit, Doc Suza and Doc Vang, for a total of five. I asked Captain Gratton for some additional help and was given two more corpsmen from the dam. Doc Wold and Doc Bush.

I had developed a close relationship with both of those corpsmen during my tour. At different times, both of them had been assigned to me personally to correct disciplinary issues. Doc Bush had a problem waking up for missions on time and was generally late for everything. One time he actually fell asleep inside of a cooler used to store food for the kitchen. He accidentally locked himself inside and slept there for twenty-four straight hours. When the morning shift opened the door, he was almost frozen solid. We laid him out in the sun for most of the day just to thaw him out.

When I grilled the five-foot-two, one hundred and eighty-five-pound bowling ball of a Doc about what was causing him to be so late, he simply said, "I'm just tired I guess." Tired? Easily corrected. I placed Bush on mandatory sleep for three straight days. He could only get up to use the bathroom. No reading, no talking, no nothing. Just sleeping for seventy-two long hours. When it was mealtime, I brought him his food, breakfast in bed style. When people were talking and joking in the squad bay, we ignored him. When casualties came in, he slept. After treating a round of casualties, I asked Bush if he was tired of letting everyone else fight the war for him while he slept. With tears in his eyes, and a newfound resolve, he answered convincingly, "Yes, HM2."

Now, I would like to say that I was brilliant in curing the young corpsmen of his irresponsibility, but I wasn't. He was still the last one to arrive for about everything. But at least he showed up and wasn't sleeping through whole patrols. Baby steps, I guess. Bush did have one redeeming quality. He had a head like a cement block. Nothing could hurt this guy. He had been blown up by at least three IEDs to this point and wasn't shaken at all. In fact, the minute something exploded, the guy turned into a war hero, running around, shooting and treating patients like this was his third tour. Apparently, all he needed was someone to throw hand grenades at him his whole life.

Wold on the other hand, was a different story. Thor Wold, a six-foot, two-hundred-and-twenty pound, Scandinavian polar bear was just, well…a jerk. His attitude was terrible, he was lazy, and he had no respect for authority whatsoever. All in all, he was a mess. The upside to Wold; he was smart. Smart like med school would have bored him to death smart. Smarter than I was. And his medical skills were impeccable. I tried and tried to make young Thor into a model corpsman, but I failed. He pissed people off just as bad the last day he was in Iraq as the first.

I would have thought that giving these two corpsmen to me to take to the front lines would have been enough to send me over the edge. Quite the contrary. I was thrilled. Bush and Wold didn't care for any part of the military. Not the rank structure, not the discipline, not the uniform or the camaraderie. The only thing they cared about was taking care of casualties. Marines, civilians, insurgents, Iraqi soldiers, it didn't matter. They liked practicing medicine and saving lives. Who wouldn't want them on their team? Heck, I loved these guys.

Wold, Bush, and I shipped out the next day to meet India at Al-Asad. It was nice to be back with India. They were my boys. For almost two years, I had drilled with them in Buffalo and knew every one of them. It was like coming home again…but much more awkward.

When I showed up at Al-Asad, I was filthy. I had about two days' growth of facial hair, my pants were ripped at the knees, and my blouse which I had just recently replaced because it was in such terrible condition, had no name tapes on it. Coming into a base that was very spit and polished, it was about two minutes before someone yelled at me. "Hey Devil Dog! Where the fuck are your name tapes?" I had no idea who the guy was yelling at me, or what his rank was. I just barked out a quick, "Aye-aye," and kept walking, pretty much ignoring someone who looked like they hadn't missed a meal or an hour's sleep for the whole war. I didn't have the energy or patience to be yelled at by a guy who ate Burger King while I was eating cold chili mac from an MRE.

Wold looked at me and didn't miss the opportunity, "And I'm the one who doesn't show respect for authority?"

"Eat a bag of dicks, Wold."

"Cleveland, you should have just told him that your name tags fell off when you got shot in the head." I had to give it to him, the big hairy Viking always made me laugh.

Assimilating back into India Company was a bit of a challenge. Everyone knew about Mother's Day, and everyone missed the hell out of Jeff. What bothered me the most though was that the Marines of India started telling each other ghost stories about that day. Stories that made me out to be some type of hero. Some of the Marines from second platoon asked me "Doc, is it true that you're getting put in for the Silver Star?"

I couldn't even look them in the face when I told them, "No." I was ashamed of myself that an entire company of my friends thought I was some type of hero, when I clearly wasn't. I hadn't been able to make it to my best friend in time to even try and save his life. Failure is not heroism.

I was also concerned that these boys had no idea what they were getting into. None of them had seen combat nor had any idea what it was going to be like.

Even though I felt bad for them stuck back on guard duty, I was very close to those Marines, and I didn't want them to have to go through what I had been through.

"You think were gonna see some shit, Doc?" The eagerness in their voices reminded me of how everyone acted when we were training in Twentynine Palms, before we came to Iraq and started getting all shot up.

"Boys, I hope to God all we do is sit in a firm base and wait until September, cause I want all of you to come home in one piece..." and because I couldn't live with myself if they didn't. I paused for a bit, trying to swallow the lump in my throat before I continued.

"But," I continued, "I guarantee I will be calling for a casualty evac within the first forty-eight hours."

I could hear every one of their hearts beating. None of them thought I was kidding or exaggerating to make a point. That was good, because I wasn't. Someone was getting shot within the first two days we operated in the city. As I walked away, I am sure they were all wondering who it was going to be.

Within a couple of days, India Company had loaded onto a series of truck convoys from Al-Asad, south to FOB Hit, the logistics base that had been under near constant mortar fire for months. Lima and Kilo companies had finished clearing the city and were running patrols around the clock. With the city under US control, the mortars stopped almost completely.

When we pulled into FOB Hit, Marines were emerging from the inside of buildings like mole people, who had been living underground for months. Their living conditions were terrible. Marines were showering with their flak jackets on, or not showering at all. The porta-johns all had shrapnel holes in them, and there were random pieces of equipment that I just couldn't identify, that had been blown up, just lying around. The dam in Haditha was crappy, but this place was downright grungy. We spent the night here before the next day's movement into the city of Hit, that would be our home for the remainder of our tour.

In the city, the bulk of our company, and about a hundred Iraqi Army soldiers would occupy a vacant teacher's college on the edge of town, that had been turned it into our base of operations by Lima Company. This was the first time that I had performed joint military operations with the "*jundis*," Arabic for "soldiers." Back at the dam, an Iraqi Army battalion had been placed in a tent city just outside of the base. At the time, I thought it was odd that the powers that be would make an entire Iraqi battalion live outside the safety and security provided by an American combat outpost. But within a week, I could see why that decision was made. I don't mean to be disparaging in describing the army of our host country. They were doing the best they could under the circumstances. But to put it mildly…they were a pain in the ass. It was like trying to fight a war with a boat anchor tied around our necks. Instead of being a force multiplier, they were one more thing added to the list of crap we had to worry about on a daily basis.

Officially, we were tasked with teaching them how to fight a war for themselves so they could eventually lead the fight against Islamic extremism in Iraq and the US military would be able to return home, victorious. I want to say that I agree wholeheartedly with that policy, but in practice, it was excruciatingly slow and painful.

To the creators of this policy, I would like to point out a couple of personal issues that I had with it. First, and most importantly, I didn't know how to fight a war. This was my first one. What I knew, I learned from training. And let me tell ya, pointing your rifle at a fellow Marine pretending to be the enemy, in Twentynine Palms, California, saying "bang" and watching him play dead, was a heck of a lot different than pointing your rifle at a man in Iraq who may or may not be the enemy, pulling the trigger, and having him die. Hard to teach that lesson to someone through power point or interactive discussions.

In fact, this was the first war for most of us. We were all just trying to figure out how to stay alive and kill some bad guys. By the end of our tour, when we had more than a few hard-fought lessons to pass on, we were so spent that we weren't very good teachers.

Second, it was hard to teach someone to fight a war while fighting it. For example, a patrol of just some Marines and some *jundis* on a little joint exercise wind up ambushed. In a perfect teaching scenario, this was what would happen.

The Marine corporal would hear the explosion from an IED, see his point man vaporized in the blast, and immediately come to this conclusion "my, what a marvelous teaching moment for these young Iraqi warriors." The twenty-one-year-old corporal would then walk up to the seventeen-year-old Iraqi Army squad leader, all while small arms fire snapped past them and would take a moment to explain the counter ambush principles that they had practiced. "Ahmed… you see what just happened? Lance Corporal Haag was killed by an

IED, and some bad guys are shooting at us from on top of that roof. Now what you need to do is; call this in on the radio, have some of your guys shoot back at the guys on the roof and have the rest of your guys go around to the side of the house with the bad guys on it and flank them. After those guys on the roof are killed, then we will go door to door and try and find the guy who set off the IED. Any questions?"

A bit impossible to say the least.

What actually happened was this: the very same patrol was walking down the street. An IED was detonated and Lance Corporal Haag, walking as the point man went down, but was not killed. He was missing an arm and a leg and begging for help. Because the rifle squad was all composed of reservist Marines from northwestern New York, Haag also happened to be the cousin of the radio operator, and the SAW gunner's best friend since third grade, both of whom started crying hysterically and instantly ran to help Haag. A Marine towards the end of the patrol saw what he believed to be an insurgent spotter who may have triggered the IED and shot at him. The *jundis*, seeing this horrific event and hearing the gunfire, decided to shoot also. Only they didn't know where to shoot so they did what was called the "death blossom," meaning all of the *jundis* shoot all of their ammunition on full automatic in every direction, spraying vegetable carts, meat shops, houses and people, with automatic weapons fire, none of which needed to be shot.

The corporal, who also happened to go to the prom with Haag' sister, had to deal with three major problems. First, he ran to Haag and made sure that he was taken care of, and that a medevac was being called in to help. Next, he ran to the rifleman who was shooting at the spotter and directed more Marines to fire on the same position, because now the spotter was shooting back at the Marines and directed a second group of Marines to maneuver on the insurgent as soon as the *jundis* stopped shooting. Then, the corporal ran to the Iraqi squad leader and told him to get his guys to stop shooting, only the Iraqi squad leader didn't speak English, so the two yelled back and forth until the *jundis* just ran out of ammo.

By then, the Marines who engaged the spotter would most likely have shot him and were advancing on the house. The *jundis*, who didn't know what the hell was going on because no one could tell them, saw the Marines running to the house. Out of ammo and not wanting to be left behind, the *jundis* ran after the Marines. Only the corporal needed the *jundis* to stay behind and protect Haag, the corpsman, and the radioman while they waited for the medevac because he didn't have any more Marines left. But, as the corporal runs after the *jundis* to tried to get them to set up security, he realized that even if he managed to herd them into place, they didn't have any bullets left to shoot bad guys, so he just let them run around like wild animals while he made sure that his future brother-in-law could still be a groomsman in his wedding. The ambush was over anyhow, two minutes and seven seconds had passed. No one was shooting anymore.

Everyone gathered around Haag, whose face was barely recognizable through the burns. The medevac took Haag to the BAS, then to Al-Asad, then to Landstuhl, Germany, and finally to Bethesda, where he was fitted for prosthetics.

The patrol returned to the forward operating base. The corporal reported one insurgent KIA, but the local town leader is demanding to meet with the company commander saying that the Marines shot two women also. The corporal is questioned and must account for rounds that were probably fired by the *jundis*.

The next day the patrol goes out again. The *jundis saw* what happened to Haag, so they don't really want to go out again, and they show up late to the patrol. The corporal, knowing what a wreck the *jundis* were the last time, steps the patrol off without them, happy to be rid of the headache. Later he explained to the company commander that the *jundis* were a no show. The company commander meets with the Iraqi Army commander, and they sort the situation out.

Then the company commander then met with all the corporals and reminds them that they are professionals, and that it is their mission to train the Iraqis and that we will be here until we accomplish that mission, and that there will not be another patrol without *jundis*. The patrol steps off again the following day, Marines and *jundis* together, with the corporal looking to make better use of a "teachable moment."

This scenario plays out day after day, the only thing different is the name of the lance corporal. During each engagement though, the *jundis* would take a smaller step towards self-sufficiency. It was a process, however, that was measured in years, not patrols. No wonder it took from 2003 until 2010 for them to get up to speed.

We arrived at the teacher's college occupied by the Lima Marines. Lima Company had lost by far the most Marines of the entire battalion. They were the workhorse. Not an operation in the entire 2nd Marine Regiment happened without Lima Company being involved. As I was getting a briefing from one of the Lima Corpsmen, HM2 Wallace, I realized just what shells of human beings the Lima Marines had become.

"How have you been staffing your patrols?" I asked him.

"QRFs (quick reaction force) and high-risk patrols get a corpsman, the rest we try and fill in where we can." Wallace didn't even look up from wringing the sweat out of his shirt as he spoke.

"Not all your patrols have Docs?" I asked, a little surprised.

Wallace looked at me with hollow eyes. "We're pretty short."

Lima had lost yet another corpsman when HM3 Youngblood was hit by an IED the day before. Travis Youngblood was an active duty corpsman from Great Lakes who had arrived with the rest of the active duty augmentees. He was a favorite of all the Great Lakes Boys. Travis was a bit older, twenty-six, and was married with a kid at home. His age, and calm disposition, made him someone that all of his friends looked up to. Travis was evacuated immediately after his injury. Although he was hurt badly, he was still talking and in good spirits when he arrived at the BAS. Travis was flown out to the field hospital at Al-Asad, and then on to a higher level of care from there. Statistically, there was well over a 90% survival rate for wounded Marines who made it to the hospital at Al-Asad, so everyone breathed a collective sigh of relief when we learned he arrived there safely. I think that's what made it so hard when six days later, we learned he died. Travis suffered an injury called "blast lung" where the air sacks in your lungs swell because of the overpressure from the blast wave caused by the IED. The swelling takes place over time and causes your lungs not to be able to exchange air. Travis's injury was not survivable from day one, we just didn't know it. Travis Youngblood was a good man.

All in all, Lima had probably less than half of their original corpsmen still in the fight. I said a quick prayer that they would be able to take it easy the rest of the deployment. They didn't. Lima would go on to have another dozen Marines killed and twice as many wounded before their war would finally end.

The accommodations at the teacher's college in Hit were nice, there was more than enough room for everyone to sleep comfortably. We had porta-johns, which was like heaven compared to the alternative, and we even paid a local contractor to empty them, which also beat the alternative.

The corpsmen crafted a makeshift aid station on the first floor of the building as our own personal living space. We were fortunate to be able to make some significant improvements to the space courtesy of a section from the 4th Engineering Battalion that was attached to us. Their OIC, Sergeant Howard Cook, made sure that the corpsmen, and our small aid station, were his first priority when it came to construction materials, sandbags and Hesco barriers for blast protections or generator power.

Sergeant Cook's improvements allowed us to set up cots to see sick Marines during the day and to sleep on at night. He built shelves to hold our medical supplies, and then stole medical supplies from other units to fill the shelves. He sandbagged the walls to keep the mortars out. He stole a cooler filled with near beer for Marines that needed to take a break. He set up a nice sound system. He made it home.

For the next four months, Sergeant Cook and his engineers would be some of our best friends. Howard was a graduate of the Virginia Military Institute, and was fascinated with medicine, and told us on more than one occasion how impressed he was with our ability to evacuate casualties. After his deployment, Howard transferred into the Army Reserves, was commissioned as an officer and trained as a Blackhawk Pilot for a Casualty Evacuation

Squadron. Howard wanted to continue to serve, but he wanted to help save lives instead of taking them. Unfortunately, Howard was killed in a car crash before he could fly his first mission. Howard Cook was a good man.

Now with all these luxury accommodations that Sergeant Cook had built us came a certain responsibility. The responsibility to share. Every amenity we owned had to somehow be of benefit to the entire one hundred and seventy-five Marines in the company. We had a coffee pot (and an outstanding supply of Dunkin Doughnuts coffee shipped in from the States), so we made sure that every morning, the officers, the Marines manning the Combat Operations center, and every Marine standing guard who wanted one, had a cup. We had a laptop, so we made sure that every Marine who wanted could share his iTunes playlists with it. We had a great supply of spices, so we would cook dinner for a squad at a time to give them a break from MREs. And we had a couch so anyone that needed could come in, lie down, and cry about their girl back home dumping them, their friend in Kilo Company being obliterated by a mortar, or about being too scared to go out on patrol again.

All of that stuff we managed to scrounge for our Aid Station might have been a little cush, but it was a real tool in keeping Marines in the fight.

CHAPTER 12
CHICKEN MAN

Our patrol schedule was insane. India Company had four platoons. Each platoon consisted of three squads of twelve to thirteen Marines. For the most part, combat operations in the city of Hit were run on the squad level. Each squad would operate with about a half dozen Iraqi Soldiers, giving us close to twenty men in each operating team. My job was to provide medical support for those twelve squads with seven corpsmen.

One corpsman would constantly be assigned to our QRF, a squad of Marines waiting in a section of Amtracks, ready to reinforce anyone who got into trouble. The tracks QRF were also our initial casualty evacuation. Nowhere in the city was there a place large enough or safe enough to land a helicopter to evacuate wounded. All of our casualties would have to be taken by ground to the FOB outside of Hit, about twelve miles away. For our corpsmen, this was a hot job, and one that I could only assign the most capable corpsmen to. For the first month, the company CO made it a standing order that I personally chaperone each medevac until the newer corpsmen got some experience and were up to speed. It didn't take long. True to my prediction, we were working on day two, an Iraqi soldier took a sniper round through the thigh. On day three, Andy Lenz was hit.

Private First Class Andrew Lenz was on patrol with his squad through the center of the city. Initially, India Company would conduct routine "presence" patrols. The purpose being to let the bad guys know we were there, and to enable the *jundis* to talk with the locals and figure out where the bad guys were, so we could plan and conduct targeted raids. In those first days though, it was a feeling out process to try and figure out what parts of the city were supportive of us, and which ones were dangerous. On that patrol, PFC Lenz's squad stumbled into the dangerous part. As they were walking, an 82mm mortar shell rigged as an IED exploded, shredding Andy's right hand, and peppering his body with shrapnel.

HM3 Matthew Coleson was manning the QRF that day when the call for a medevac came. I grabbed my gear and hopped into the track with him. Matt Coleson was the most experienced of my corpsmen. He worked in the ER at Great Lakes and took every opportunity to train to be better at his job. He took his mission seriously, he was a total pro. As Coleson and his squad ran by, I grabbed my kit and jumped in the track with him. I went over the casualty evacuation (casevac) procedures with him on the way, just to make sure that he had them straight in his head; but there was no need. He knew them by heart.

I grabbed Coleson and we both popped up through the moon roof on the Amtrack. About two blocks away I could see purple smoke, the marking signal for where the casualty collection point was located. I pointed to it and Coleson just nodded his head. I was happy to see that the Marines, in their first contact, had remembered the casevac procedures.

We pulled up and the ramp to the track dropped. Lenz was cut in about a thousand places. His right hand had a huge hole through it, and he was in a lot of pain. Lenz was hurt badly, and his war was over, but he would make it. Coleson jumped right in, doing a rapid trauma assessment, exposing the wounds and getting them packed. I gave Andy a shot of morphine to dull the pain and he started laughing almost immediately.

"What the hell are you laughing about?" I asked him, as Coleson continued to wrap his wounds.

"My girlfriend. I got a letter from her yesterday breaking up with me. "Lenz said, laughing hysterically the entire time.

"What's so funny about that?" I asked.

"I got her now. When she hears I got hit, she's gonna feel like shit and come crawling back. It's awesome. She will literally never be able to break up with me. If she even thought of it, she would remember today and feel way too guilty to go through with it. I'm set for fucking life man. I'm king of the fucking world." Andrew Lenz is a good man; a weird man, but a good one, nonetheless.

If you were not a corpsman assigned to QRF, then you could be one of the three corpsmen that were sent with various squads throughout the city to maintain strategic Combat Outposts (COPs). That left three corpsmen to cover the patrol schedule. There were two daytime patrols and two nighttime patrols every day. That meant that one of the three corpsmen had to double up and patrol for twenty-four hours straight. I rotated my guys through the assignments: patrol, COP, and QRF, a week at a time. This gave the guys a rest when they were at the COP after they came off patrols, and equal time treating casualties when they were on QRF. The only person that didn't get a chance to rotate was me. I was constantly on the patrol schedule. And when I was on my rest cycle, if a casualty came in, I was on the casevac. I was never more tired in my entire life.

The disparity in food offered in Iraq is amazing. Back at Al-Asad, they actually had a dining facility that served lobster on Fridays, along with a Burger King, a Starbucks (with tremendous mocha Frappuccinos), and a Pizza Hut. The food was so good on the air base that there were personal trainers offered at the gyms to help servicemen and women in the rear who enjoyed the cuisine a bit too much, get back in shape.

The entrees at the FOB that housed our battalion command staff, the dam, and FOB Hit, were about ten steps down from that. Hot food was offered for breakfast and dinner, but it was on a five meal rotational basis so you got the same gross offerings once every week or twice a week, if the logistics convoy didn't make it through.

As truly awful as this food was, it was hot, it was consistent, and you could eat as much as you wanted.

The line companies however (Lima, Kilo, and India) survived entirely on either MREs, or plan B. Plan B being care packages sent from home or whatever people could manage to smuggle back from Al-Asad on the occasional trip to the rear. Which was fine, but I could only survive so long on tuna, Oreos, and salt and vinegar Pringles.

The Marines in the line companies were hungry. Always. Whenever we went back to Al-Asad, either to evacuate a casualty or a detainee, we would descend on the dining facility (DFAC) like a flock of vultures. A dozen smelly, nasty Marines with the knees ripped out of their pants and blood on their boots would park a row of Humvees outside of the DFAC and push in front of anyone in line who was not an officer or did not have a week's growth of facial hair. It would take two or three passes through the buffet for each of us to gorge ourselves to a point just before we exploded. As full as I was for the trip back to the line, I was always hungry again in a matter of hours.

Because the line companies were stuck in a sandy hell with no food to speak of, the Marine Corps came up with an idea of a "Mobile PX." It was a semi-truck with the trailer converted into a little store where we could buy everything from tuna packets to ramen noodles to socks and towels. It traveled around the AO with the logistics convoy, so if we were lucky enough to be at the FOB when the mortar rounds and bottled water were being delivered, we would get a chance to shop at the PX. The only thing was no one had any money.

So instead of money, they had a guy in the "store" who would let us take an advance out on our paycheck and would give us little cardboard disks called "pogs" with a dollar value printed on the front, and a motivational picture (like an eagle or a tank) on the back. Then we took our pogs and went shopping.

We only ever got to use the mobile PX once, but boy, was it a doozie. The PX would only stay as long as it took for the logistics convoy to drop off our food, water, and ammo; so, we had to shop fast. On this trip though, as the convoy was preparing to leave, we took a barrage of a half dozen or so mortar rounds that were right on target. Thankfully, the only thing of any significance that was hit was the tractor hauling the mobile PX. It wasn't fixable. They would have to leave it behind and pick it up another time. It took us about a week, but we ate every single crumb inside of that trailer.

One of our more ingenious corpsmen managed to scrounge a rice cooker from one of the local Iraqis. So, for a treat, we purchased groceries when we were on patrol through the market in town. Some rice, spices, vegetables, and flat bread and we were in business. A fresh locally grown feast was just what the doctor ordered to break the monotony between MREs and care package food.

As good as the local food was, after a few weeks' worth of trips to the market, we noticed something was missing. That something was meat. Although there was no shortage of freshly butchered goats, or day-old goat organs for sale in the market, we didn't have any practical way to cook them.

On one patrol, I noticed a restaurant in town that sold whole roasted chickens, cooked just like KFC, on a rotisserie outside their shop. They looked delicious and they smelled even better. The only problem was the owners of the restaurant didn't like Americans, so they refused to sell us any chickens. Patrol after patrol I tried, but no luck.

The other corpsmen knew of the chicken stand as well and were very interested in seeing how I was making out. One day, I was performing minor surgery in the aid station in front of a rather good sized group of Marines. For some reason, Marines loved seeing me cut open puss filled wounds, it fascinated them. I was happy to let them look. The only thing I required is that they didn't get what sterile equipment I had dirty; and that they hummed the theme to Top Gun while I was cutting. (It made me feel very Navy like.)

During the surgery, another corpsman asked how my most recent attempted chicken purchase had gone on my morning patrol.

"I tried to have some of the *jundis* buy a few, but it didn't go so well." I said, with more than a little disappointment in my voice. "Before we could even get to the market the *jundis* saw someone who they thought was *mujahideen*, so they grabbed him up and beat him with a shoe," I explained as I squeezed puss from the shoulder of a machine gunner.

"What did the guy do?" the corpsman asked, mildly intrigued.

"Hell if I know. They let him go. I asked what was going on and all the *jundi* said was, 'No problem,' and sent the guy on his way." The shop closed when they saw the beating and we couldn't get the chickens.

Just then, Corporal Vancherri from second squad chimed in, "Doc, if you tell me where that chicken place is, I will bring you back a fucking chicken, you can believe that."

Now the secret was out. If I didn't get a chicken first, the Marines would certainly get one for me, by force if necessary. God love Marine infantry, but they were not interested in starting ongoing commerce with the locals, they would just as soon pillage the whole shop for one chicken. I had to get one on the next patrol to keep the Marines from occupying it as a base of operations and ruining future chicken dinners for everyone.

As part of our mission to train the Iraqi Army, we would take their officers on patrol with us so they could speak with the citizens and enact some type of dialogue between us and the population. The restaurant was just such a place that could use some of that "dialogue." The Iraqi Army colonel in charge of our company of *jundis* set up a meeting with some local sheiks, who he met for tea inside of the establishment. My job was to stand watch over the meeting outside of the restaurant...right next to the guy who sold those heavenly chickens.

While I was on guard duty, I walked up to the chicken man, stared straight at him with my best no nonsense look, placed a $20 bill on the counter (which was like a million Iraqi dollars) and pointed at a chicken. Chicken man ignored me and helped several other customers instead, but I didn't move. Even when the meeting had broken up and the patrol was departing, I stood firm. I was not leaving without a chicken. After what seemed like twenty minutes, the chicken man broke down and gave me my bird, but not before throwing the $20 bill back at me. I thanked him calmly in Arabic, picked up the money, placed it on the counter, and walked away.

The meeting, it turned out, was productive. The Iraqi colonel addressed some concerns, and the chicken man got some business. That one transaction paved the way for my corpsmen to engage in at will chicken purchases whenever they patrolled past the restaurant. Life was good...until the Marines messed it up for the rest of us.

About a month into our occupation of the city, I was on patrol in the market with my sights set on an eight-pound roaster, when I heard our quick reaction force call over the radio. An Army Special Forces (SF) unit had taken casualties across the Euphrates River, and they were calling for a QRF and a casevac. During our deployment, our battalion had become the most actively engaged combat unit at that time, in the country. Because there was so much work, we were constantly getting new supporting units in our AO to lend us a hand. A Navy Explosive Ordnance Disposal (EOD) team to augment our Marine EOD team, a section of Naval Construction Engineers (SEABES) to help fortify our positions, an Army SF Operational Detachment Alpha (ODA) Team to help with direct action missions on high value targets, a Marine ANGLCO to call in close air support missions...the list was lengthy.

At this point, there were so many cats and dogs operating in our AO, that our patrol briefings included a very detailed section concerning where the other "blue" or friendly forces were, what frequency they were operating on, and what their mission was. For this patrol, we were briefed that Lucky 45, the SF ODA Team, would be operating on the east side of the Euphrates River. The city of Hit was home to the only bridge between the Haditha dam and Ramadi that could hold the weight of light armored vehicles. To maintain control of that bridge, a COP was established in a hotel on the river that overlooked the bridge. The hotel was pink, so we naturally called the COP "the pink hotel."

I heard the call by Lucky 45 over the company frequency for the QRF. It was midday, and our patrol route had taken us into the middle of the market, which was about a mile from the bridge that Lucky 45 was on the other side of. The road that ran through the market was the only access from the bridge to the main service road that led to the BAS.

At that time of day, the market was filled with thousands of people, carts, cars, and goats, and no one was moving. Our patrol was at a standstill.

About five minutes after the call went out, the Amtracks came screaming down the road into the market, and ran smack into a wall of humanity, who were not inclined to move for the armored vehicles.

Now the whole mission was at a standstill.

I called up the patrol leader of the QRF squad as we walked next to the tracks and asked who the corpsman was that he brought to take care of the wounded. I wanted to make sure that whoever the corpsman was, he had the equipment he needed and that he was clear on the medevac procedures. The patrol leader instantly looked aghast and said, "We forgot to bring one."

I exploded. "You're shitting me! You went on a medevac mission without a fucking medic?" That was classic Marine Corps, home to some of the toughest, and dumbest, people on earth.

So, here was the situation. We had a bunch of SF guys who needed to be rescued, and a bunch of armored vehicles that were sent to rescue them stuck in the middle of a bazaar. And even if the two managed to meet, all they could do was hope the poor guy who was shot didn't die, because the guys on the Amtracks didn't bring along anyone who knew anything about fixing up shot people.

I was furious. As the chief corpsman, it was my job to make sure that medevacs such as this went smoothly, and this was anything but smooth. I heard the SF team calling again on the radio, frantically asking where their extraction team was. I needed to get things moving and in a hurry.

So, I walked right in front of the lead Amtrack, smack dab into the middle of the bazaar, and started yelling like a mad man. I screamed in English and Arabic for people to move out of the way, and when they didn't listen, I moved them. I kicked goats. I had people turn cars around. I flipped over vegetable carts. I was running through the middle of the street like a lunatic clearing a path for the two Amtracks that were following behind me.

At one point, a taxi stopped right in the middle of the road. The driver got out and tried to run, probably scared that the Amtrack was going to run over his car. I managed to grab him as he was running and throw him at gunpoint back into his cab in sort of a reverse carjacking. The taxi driver moved his car in a hurry and cleared the road. But not before one of the Amtrack drivers veered off the road towards the chicken man's restaurant and drove directly over…the rotisserie.

Weeks of goodwill and good chicken were crushed like a pop can under the weight of the massive tank. All the things we did to try and show the chicken man that Americans were good people, and we went and crushed his livelihood. I wondered how hard it was to get a replacement rotisserie in the desert.

I continued to run the mile through the market in front of the tracks, clearing the way. When we arrived at the bridge, I met the SF medic who was treating the wounded man, who was an Iraqi. He had placed a breathing tube in his throat and was using a bag to breathe for him. The man had been shot in the chest and the wounds had been sealed.

As we carried the Iraqi into the back of the track, I started my trauma assessment. Not knowing who the Iraqi was, or if he was working with the ODA team, I asked the SF medic, "Is he a good guy or a bad guy?"

The SF medic looked at me and shrugged his shoulders. That's how it goes. There had been a scrape, somehow that guy got shot by someone. We didn't know who he was. As the track sped back through the market towards the main service road, I grabbed the radio handset and called my casualty report to the BAS.

"Bulldog, Bulldog, Echo-Five-Charlie. Caserep." Bulldog was the call sign for our battalion, and Echo-Five-Charlie was my individual call sign that consisted of my rank (E-5, hence Echo Five) and the first initial of my last name. (Cleveland, so, Charlie.)

"This is Bulldog, send it," came the reply.

"I have one urgent surgical. Gunshot wound to the chest.

Request immediate helo evac." Casualties were divided into four categories: routine, which did not require evacuation; priority, which required evacuation but was not life threatening; urgent surgical, the person needed surgery immediately or he would die; and expectant, the person had or was going to die.

Bulldog responded just like he had countless times before. "Understood, Echo-Five-Charlie. One urgent surgical. GSW to the chest. Air is enroute."

I returned to my assessment of the Iraqi and saw that he had turned gray. I checked for a pulse and there was none. "How long have you been bagging him?" I asked the SF medic.

"About twenty minutes."

I started laying out our treatment options for the SF medic. "Hey bro, it's your patient and your call, but this guy's gone. I'll work him until you tell me to stop, but we're not bringing him back." The SF medic nodded in agreement.

I updated the BAS with the news. "Bulldog, Bulldog, Echo-Five-Charlie. Update on the caserep. Now one Iraqi expectant. You can cancel the helo evac." I made sure to say long, slow, and loud the word "Iraqi" as not to otherwise upset any of the hundreds of Marines listening on the battalion frequency who would immediately wonder if it was their friend who had been killed.

The call came back. "Echo-Five-Charlie, this is Bulldog...interrogative, (a term used to specify that the next sentence is going to be a question that requires an answer) is the KIA friendly or enemy?"

I couldn't shrug my shoulders over the radio, so I gave the best answer I could. "Bulldog, that's unclear to me. Lucky 45 has the specifics." It was better to let the SF guys explain what had happened.

The ramp dropped, we carried the dead man out, and laid him in the sand outside the BAS. I shook the SF medic's hand without saying a word and climbed back into the track to head back to the FOB. I had no idea who the dead man was, or why he died. I just knew it was Tuesday, and that was pretty normal for a Tuesday.

CHAPTER 13
SEPTEMBER 4TH

Life in the city of Hit had settled into a routine for our rifle company. Albeit a gruesome and surreal routine, but a routine, nonetheless. We patrolled the city on pretty much a constant basis. At any given time, there was usually a squad of Marines and a squad of Iraqi Army *jundis* walking through the streets, trying to suppress evil.

Each day passed pretty much the same. Patrol through the city, sweat, come back, eat crappy food, watch a pirated DVD. Patrol through the city again, sweat, come back, eat crappy food, watch a pirated DVD.

The only break in the monotony was when the patrol was ambushed or blown up by an IED. That wasn't an everyday occurrence, but it happened with enough regularity that I had grown used to putting the pieces of my friends back together, attending their makeshift memorials, and crying myself to sleep over it. That, too, was part of the routine.

Our routine was interrupted in spectacular fashion on August 4th. After returning from our patrols, myself and three of my corpsmen, Matt Coleson, Thor Wold, and Troy Suza, were taking a much-needed mid-morning nap in our makeshift aid station.

Each of the three corpsmen was active duty Navy who had been assigned to our reserve unit prior to our deployment to augment our medical staff. Although none would admit it, I am sure that they were all less than thrilled that they were plucked out of a Naval hospital in Illinois and sent with a bunch of old reservists to a real shooting war. But the truth be told, all three were great medical providers. Calm and quiet Coleson, always ready and able to help, never required a minute of supervision. Wold and Suza, on the other hand, required a bit more supervision.

Both of them, at different times, were assigned to me for corrective action. Wold had a bit of a propensity to question authority, so for a week he was assigned as my shadow. He had to walk wherever I walked, eat whatever I ate, run however far I ran, all in an effort to make him a more disciplined sailor. It didn't work a bit…and that made me like him even more. Suza needed remedial help with his Marine Corps skills. He was terrible at patrolling, and could care less where his rifle was, or what he might need to shoot with it. I actually walked behind him on one patrol and threw rocks at him every time he did something wrong. That didn't work a bit either. Suza cared more about his medical work and playing guitar… and I liked him more for that, too.

Our nap was unfortunately cut short by small arms fire coming from some of the outside security positions. Gunshots were nothing new, but it did make it hard to sleep.

What was new, however, was that the gunshots were not subsiding. Just as I took note that the volume of fire was not decreasing, I heard one of our sniper positions on the roof start engaging targets, which had very serious implications. I stood up with the intention of getting dressed and reporting to the Combat Operations Center to get up to speed on what all the shooting was about. I got one boot half on when our room vaporized around me in a massive explosion. A wave of debris from the blast overtook me and threw me against the only wall left standing, knocking me unconscious.

As I started to awaken, my ears were burning; my head was pounding, and my left leg felt like I had broken my hip. I could barely move and felt like I had been seriously wounded. I tried to yell out for help but could only moan. My mouth just wouldn't form words. I couldn't breathe.

Lying face down, I could barely see through the smoke and concrete dust that enveloped what used to be the BAS. What I could see was that Marines were laying everywhere, and any Marine that was not lying wounded was running around shooting. RPGs and AK-47 fire slammed into the building. We were quite obviously under a significant attack.

I couldn't move. I couldn't talk. I couldn't think. Our building had been blown up and now all the Marines, who were able, were fending off what sounded like every insurgent within a hundred-mile area.

Bodies lay everywhere, Marines were shooting, people were screaming. I looked over to where Coleson had been sleeping and it was a mountain of rubble. I remember saying to myself, "Coleson is dead."

All I could manage to do was crawl under what was left of my cot, say a prayer, and bleed until I could catch my breath and figure out where I was. It really wasn't much of a prayer. It consisted mostly of me saying repeatedly, "Please God, take care of me. Please God, take care of me."

After about the fourth time saying it, I could hear screaming from under a pile of rubble in the room. It was Coleson. He had been buried under what was the wall of the BAS and was yelling for help. Coleson's cries served to snap me to attention. I was hurt but not dead, and I needed to quit hiding under a bed and do my job.

As I stood up, I could see a piece of wood from what had been our medical supply cabinet sticking out of my leg. I managed to yank it out and crawl over to Coleson, who was trapped under a large pile of concrete. I yelled for help from anyone who could hear me, "Two down in the BAS!" The other two corpsmen that had been in the BAS when it exploded, Troy Suza and Thor Wold, came limping in from the hallway to help. Between the three of us, we managed to pull Coleson from the wreckage.

Each of them looked as bad as I felt. Wold had been knocked unconscious as well and was barely able to stand and Coleson almost certainly had a broken leg. Suza had a nasty gaping shrapnel wound to his arm and was yelling, "Morphine! Morphine! Someone get me some Morphine!"

I asked Suza who the morphine was for, and he replied rather sheepishly, "It's for me, my arm hurts real bad and I think if I have some morphine, I might be able to do a better job." Wold answered before I could, "Troy, you're an idiot, you can't have morphine."

With the four of us now decidedly not dead, we needed to get down to the business of treating the casualties that were piling up around us. One of the Marines whom we had given extra medical training approached us and asked us what he could do to help. I gave him an empty cardboard box and instructed him to sift through the rubble of our aid station and find anything that remotely looked like a usable medical supply. To add insult to our collective injuries, all our medical equipment had been destroyed in the blast, and I needed the Marine to start finding us something to work with.

Next, I set up areas for triaging and treating Marines. Each wounded Marine would be classified in order of the seriousness of his wounds, with the most severe being treated first. I gathered Coleson, Suza, and Wold in a huddle, "Anyone who is hurt but can still walk, goes in first platoon's berthing and we'll get to them later. Anyone who is hurt seriously bad goes into the interpreter's berthing and you guys work on them now. Anyone who is dead or going to die goes by the steps. I'm going to start rounding up hurt people and bring them back to you." Sounded like as good a plan as any.

I took two steps before I was stopped by an Iraqi Army soldier who was covered in blood and screaming. "Many *jundis* hurt! Please Help." Our base consisted of two separate buildings in the same compound, one for US Marines and one for Iraqi Army soldiers. Each force was responsible for fortifying their respective buildings to help withstand an attack. The Marines, Sergeant Howard Cook in particular, did an exceptional job of filling sandbags and making things a bit safer. The *jundis*, however, were lazy. They had made no such fortifications and were now getting slaughtered in the attack.

I grabbed Sergeant Cusmano, who was now coordinating the Marines on the first floor. The courtyard separating the two buildings in the compound was about a hundred meters wide and there was a good amount of shooting going on which made crossing it to get to the *jundis* more than a bit tricky. I asked Sergeant Cusmano and his Marines to provide covering fire to allow me to do it a little more safely. The Marines didn't need a whole lot of encouragement to shoot at stuff and were more than happy to put down a good body of covering fire.

Sergeant Cusmano, however, did one better. "I'll go with ya, Doc." As crappy a job as it is to go running across an open field wearing only boots, boxer shorts, and a flak jacket, all while getting shot at…it was still my job. It wasn't Sergeant Cusmano's. He could have easily stayed behind and been much safer, and I never would have thought less of him for doing so. But he didn't. Rob Cusmano was a short, tough corrections officer from New York City. He was one of the best squad leaders in the company, and he never complained...about anything. Cusmano had been an active duty Marine who had ended his service before being able to fight in the war. He joined back up for the sole purpose of getting into the fight. Rob Cusmano is a good man.

We sprinted the hundred meters, shooting as we ran. We burst into the building, alive and out of breath, only to find the bodies of about a dozen critically injured *jundis* littering the floors of the Iraqi Army post. All of them needed a helicopter and surgery immediately. One *jundi* was crying and pointing to a lifeless body lying in the street.

I looked quickly for the Iraqi Physician's Assistant who served as the chief medical officer for the unit. I found him…with a hole in his chest the size of my fist. The injuries to the *jundis* were so massive that there was nothing I could do to treat them where they lay. We had to get them back to the triage points.

Sergeant Cusmano, some uninjured Iraqi soldiers, and I grabbed the worst of the wounded, picked them up, and ran them back across the hundred-yard dash under the continued covering fire of the Marines.

We brought the soldiers right into the room designated for the most critically wounded, and I told Wold that there would be many more coming his way. The trip across the courtyard was a killer. My leg was killing me. I could barely walk let alone run and carrying a wounded soldier while doing so had me whipped. I looked to Sergeant Cusmano, who was in much better shape than I was, and saw that he was covered in blood and sucking wind as well.

We looked at each other, gave a quick nod, and ran back to the Iraqi building to get more wounded. Sergeant Cusmano and I made four more trips across the courtyard, and it took us about twenty minutes to finally carry all the dead and wounded Iraqi soldiers to our triage point. Thankfully, neither of us was shot while doing so.

People talk a lot about heroes. While a lot of the things we did in Iraq may have seemed heroic, really, we were just doing our jobs. I was doing my job by going to help the *jundis* and Sergeant Cusmano was doing his job by providing me with the covering fire to do so. Nothing heroic about that, it was just work. Heroism is risking your own life to do something that is beyond your job, like running alongside a corpsman that is going to treat wounded soldiers when you could have stayed back and played it safe. He didn't have to go with me, but he didn't hesitate for one minute in deciding to do so. That was heroic.

By the time all the Iraqi wounded had been gathered, Wold, Suza, and Coleson had triaged and treated most of the wounded Marines. Thankfully, none of the Marine injuries were fatal. We had many, many wounded Marines. But as soon as we herded the walking wounded into the triage area, they would leave to go pick up their rifles, and get back in the fight.

I would venture to guess 75% of the Marines in India Company had a traumatic head injury. Andy Enser was wandering aimlessly down a hallway; his bell was rung so good he couldn't even walk straight. But when I tried to get him into the triage area, he refused to go. So, I set him and his weapon in a window, pointed him towards the bad guys, and let him go to work.

I checked on the trio of corpsmen who were working feverishly to save the lives of their Iraqi counterparts.

I walked into the room and was awestruck by the way those corpsmen worked. I wished the world could have seen my brave men. As I am writing this, I have the image burned into my brain. I wish I could have taken a picture of it, because writing about it is not going to do it justice. My three corpsmen, all of them wounded significantly, and none of them over twenty-two years old, looked like a team of trauma surgeons. Each of them was working on at least two patients at a time, with whatever supplies they could scoop out of the cardboard box that was at the center of the room. Periodically, the Marine I assigned to gather medical equipment from the rubble would arrive with an armload of dust covered bandages. Immediately, the fresh bandages would be swarmed on, torn open, and thrust into some wound to stop bleeding.

Suza and Coleson worked as one to treat their patients. Suza only had one usable arm, so Coleson helped him tie his bandages. Coleson couldn't walk, so Suza ran around the room gathering materials. Wold had complete control of the room. He looked so calm giving orders I would have thought he had been practicing medicine for about fifty years. I wished everyone could see my brave men.

I took a count of the wounded, which now numbered about thirty, and started planning for a medevac. I went to the COC to relay my evacuation plan to the watch officer. What I found in the COC was a complete mess. Not only had our building been blown up, but it appeared every USMC unit operating in the city of Hit was under attack as well. Our foot patrol in the city had been ambushed and was pinned down in a house. Vince Geckler, a thirty-year-old bald Wiccan corpsman, was in that house fighting for his life with his squad, who were surrounded by insurgents. Vince was a guy who hadn't wanted anything to do with the Marines. He loved the Navy and loved medicine, but this was not what he had signed up for. There were a lot of active duty guys from the Naval hospital that felt the same way. But what was different about Vince was that he was still 100% focused on his job. He hated being here, but that had no bearing on how he performed. Every day, he worked. Every day, he did his best. He never let his emotions affect his professionalism. Vince Geckler served bravely on September 4th and on many other days throughout his tour in Iraq. Unfortunately, when he got back from the war, he struggled with the things that he had seen. Some years after coming home, Vince lost that struggle and took his own life. Vince Geckler was a good man.

Our combat outpost protecting the bridge over the Euphrates River had come under attack. Insurgents managed to blow up the bridge that the combat outpost was set up to protect. Blowing up the bridge served to trap another patrol of Marines, who had also been ambushed, on the far side of the river. Trapped in the palm groves on the far side of the river was HN Robert Bush. One of the craziest things about being next to the Euphrates River was that while we were surrounded by desert, two hundred yards from the river looked like a jungle. When telling his story later, Rob Bush said, "It was like a Vietnam war movie. We were in the fucking jungle. Tracers flying by, people throwing hand grenades. We couldn't get out. All we could do was shoot back and call for a helicopter to come pick us up since the bridge was gone. A fucking helicopter landing in the middle of the jungle, door gunners shooting. It was like I was watching Platoon." Robert Bush served bravely on that day, and on many others. Robert Bush is a good man.

I found the company First Sergeant and started working through the casualty evacuation, which was a problem because literally every single Marine in India Company was under fire. I informed the First Sergeant that the casualty count was at about thirty, with one Iraqi Army KIA and several urgent surgical patients. I told him that we would need to spin up the Amtracks and start making relay trips to the BAS, where the helicopter evacuation point was set up. I also told the First Sergeant that he needed to tell the battalion to prepare for a mass casualty event and let them know that all the corpsmen for India Company had been wounded. "All of the docs?" The First Sergeant asked, in amazement, "All the ones that were here, the blast blew up the aid station with all of us in it. A couple of us can wait till the last trip out, but we all need to be medevaced."

With all the wounded stabilized, we started loading the worst onto the Amtracks. The first trip took the Iraqis who were clinging to life by a thread. The next trip took the most seriously injured Marines, including Coleson and Suza. The last trip took the walking wounded, who consisted mostly of Marines who were knocked unconscious or had head trauma. Ironically, a lot of this head trauma occurred when the air conditioning units that the combat engineers had just installed in our building blew out of the walls and came crashing onto the Marines sleeping below. Insurgents just hate air conditioning.

Each time the Amtracks returned from dropping off their cargo of wounded at the evacuation point, we were there to fill them with fresh bodies. Thankfully, on one of the return trips, the Amtracks brought with them fresh corpsmen from battalion to replace use. After all of the wounded were loaded and the replacement corpsmen arrived, Wold and I loaded ourselves.

By the time I arrived at the aid station, helicopters had taken the most seriously wounded to the hospital at Al-Asad. Laying in the trauma bay, bloodied and silent, were Wold, Suza, Coleson, and me. I had arrived last and was last to have my wounds dressed. The medical staff at the BAS was whipped. A combination of the physical exhaustion treating the flood of patients we sent them, and the emotional overload of hearing that their friends' sleeping quarters had been blown up with them in it had taken its toll. The look on their faces was half anguish for having to sew up their closest friends and half relief that their friends were still alive to fight another day. In a war where most infantry battalions would not lose a corpsman during a deployment, we had lost two so far, and most of us had been wounded. It was trying for all of us when one of our own was hurt. The boys at BAS were no different.

After my wounds were dressed at the BAS, I limped into the berthing area to check on Wold, Coleson, and Suza. When I found them, they were standing beside a freshly made bed, with clean sheets that had been pulled back as if they were going to tuck someone into bed. Wold looked at me and said in the gentlest voice, "It's okay. You can take a break now." The bed they had made was for me.

Those men had just been through the most violent experience of their lives. They had all been wounded, they had all saved human beings from death. They had all seen their friends bleed. And after all of that, the only thing they cared about was taking care of me. How do I tell men like that that how much I loved them?

I slept for about eight hours before I sat up and spoke with the chief on duty in the BAS. Getting wounded set in motion a chain of events within the Marine Corps. Soon after I was treated, the stateside staff for India Company back in Buffalo saw on the casualty tracker that I had been injured in combat.

A very short time after learning this, a staff NCO contacted my wife and informed her of my injury. Of course, being four thousand miles away he didn't have the exact details of what had happened or even the correct information about my status.

The phone rang in the late evening, just after Louise had fallen asleep. "Mrs. Cleveland, it's Master Sergeant Wheeland from the United States Marine Corps. I regret to inform you that your husband has been wounded in action."

Alone and terrified, she broke through her tears and asked, "Is he okay?"

"Ma'am, he's in a coma. I don't have any more details. Someone will call you back when we know more."

Louise didn't even hang the phone up. She fell to the floor in a heap and cried. Alone.

Ironically, the Marine Corps felt that if Marines were wounded, it was best for their loved ones to hear that they were okay from their own lips and not someone else's. After patients were stabilized by medical providers, staff NCOs would pass satellite phones around to the wounded warriors telling them in a gravelly voice to "call your mom and tell her you're okay." I thought that was a great idea. Not only did it help the families deal with the trauma of having their son blown up four thousand miles away, but it also made the Marine feel better as well. And sometimes it also gave people the chance to say their last goodbyes.

I had passed out before the BAS Chief had a chance to give me the satellite phone to call home. While I was sleeping, the staff called my wife to tell her that I was in a coma. I was not, and never had been, in a coma.

It was the middle of the night when I called her. A soft, half-sleeping voice answered the phone. I loved how no matter how soundly Louise had been sleeping, the affection in her voice always overshadowed her exhaustion when we spoke.

I forced out a rather hoarse, "Hey." I was a wreck. A bunch of my friends had been blown up, I had been blown up, I hurt like hell, and I just wanted someone to cry to. So, that was what I did, I started sobbing. It was a couple of minutes before I could catch my breath.

Louise was silent, like she was listening to a ghost on the other end of the phone. She managed to break through my crying and asked, "You're okay? They said that you were…"

I cut her off, "I got hurt a little bit." Silence.

Now it was her turn to cry uncontrollably. I felt ashamed of myself. I went and volunteered for this whole mess and now look at me. My best friend was dead, my body was all blown to crap, and I just sent the love of my life into a nervous breakdown. To say that I was a failure as a husband at this point in my life would indeed be an understatement.

Our conversation lasted only a couple more minutes, and Louise didn't say another word. I tried to reassure her as best I could. "I just got knocked out and a little cut…Honey…I'm okay, I promise. I'll call you when I can."

Silence.

Louise hung up and sobbed for days, not knowing whether I was fine or comatose. Louise carries the emotional scars of that phone call to this day. Years after I returned home, we were, stupidly, watching a documentary on the Discovery Channel about an Army hospital unit in Iraq. A soldier had been hit by an IED and was lying on a gurney, legs bandaged, his clothes and face covered in dried blood and soot, the guy was a mess. He was talking on a satellite phone to his wife, breaking the news to her. "I'm fine honey. I just got my bell rung a little bit…it's just a cut…I'll be out of the hospital in days. I'm fine…I promise." The guy was clearly lying. He was lucky to be alive.

Louise looked at me like she saw a ghost. Three words came out of her mouth so slowly, I could see her making the connection as she spoke. "That…was…you."

I nodded my head softly. It was me. Not literally, but the wounds, the words, the expression…it was as much me as it was that soldier on television.

Louise ran crying from the room. For the next week it was like August of 2005 again and she didn't know if her husband was alive or dead in a ditch. I spent the next few days reassuring her that I was home, alive, and okay. In many ways, I am still doing that today.

In all, my corpsmen treated about thirty patients, many with injuries so severe, that they would have certainly died without immediate aid. And other than the Iraqi that was killed instantly, everyone survived. They saved a dozen lives, under fire, with no functioning medical supplies, in a blown-up building, all while ignoring their own injuries. None of my corpsmen got a medal for what they did that day. No one of a higher rank than me ever told them how well they performed. None of the wounded ever thanked them. They never got the recognition they deserved for performing so flawlessly.

That was part of the reason I wanted to write this book. To tell them what an absolute honor and privilege it was to work with them. To tell them how in awe I was of how they acted that August day, four thousand miles from home. And to let everyone else know about it as well. Matt Coleson, Troy Suza, and Thor Wold are good men, and I love them to this day.

Another part of the reason I wanted to write this book is Chris Marchewka.

That day, September 4, 2005, a guy very few have ever heard of, in a city in Iraq that almost no one has ever been to, saved the lives of over a hundred Marines. Not many people have ever heard of him because he didn't get a medal for what he did. There was no documentary about him. Many people never even thanked him. That day, on the roof of our firm base, Lance Corporal Chris "Chewy" Marchewka shot the driver of the truck carrying over two thousand pounds of explosives just before it struck the building where we lived. The truck detonated at the gate instead of impacting the building. And although many of us were hurt, some very badly, we all lived. Today, because of Chewy, there are almost a hundred wives who still have husbands. There are two hundred parents who did not have to receive a knock on the door telling them their son was killed in action, and there are three hundred children who have dads. I have a wonderful life today thanks to Chewy and I want to take this opportunity to tell him how much each day means to me. I would tell him in person, but I can't. Chewy died in a motorcycle crash shortly after returning from Iraq. So, the only way I can thank him is to make sure that everyone I know knows who he was and what he did. Chris Marchewka was a good man.

CHAPTER 14
HIT 2

The comfort that routine had provided us at the beginning of our occupation had turned toxic. There had been a sense of security in sleeping in a familiar cot, patrolling a familiar street, and even eating the same familiar crappy food. That was the known. What we had feared was the unknown. The ambush. The IED. As the days turned into weeks, the mortars, ambushes, IEDs and the casualties that came with them, had ceased to be unknowns, and were now certainties.

At first, I could live with the fact that we would suffer only a handful of casualties in a week. In a company of almost a hundred and seventy-five, with about a hundred and fifty *jundis*, those were pretty good odds. But as time dragged on, I realized that it wasn't a handful of casualties a week, it was three or four more of my friends *every* week. It never stopped. Sooner or later, someone close to me just wasn't there. Sooner or later, I wondered if I would be here.

The patrol schedule, combined with the constant flow of casualties, was a grind. It ground away the very fabric of who I was. One day I was putting an IV in Jimmy Caflish after he got shot by a sniper, the next day I was putting Sergeant Pope's arm back together after it was shredded by an IED. And then Buck Byers had his arm and leg blown off by a mine and I couldn't even get to him. All I could do was run after the Humvee and scream directions at the corpsman as it tore off towards the BAS.

And those were the guys I was there for; those were the guys that I could help. I had friends in the battalion operating all over the AO that were getting hit and I was powerless to do anything about it. A week goes by, and another Lima track hits a mine and fifteen more of your friends were evaporated. Two sniper teams were overrun, and six Marines killed, including Brian Montgomery and Jeff Boscovich, who the bad guys had kidnapped and tortured. And on and on and on. It never stopped.

When all the corpsmen for the battalion were in Camp Lejeune before deployment, doing some administrative work, someone took a picture of a group of us in our berthing area. In the picture was me, Jeff, Jimmy Alunni, and Vic Lewis. I thought about those guys often. They were my boys.

Jimmy got hit by a car bomb and had a chunk of shrapnel half the size of your fist blown through his neck and another through his lungs. He was hurt so badly there wasn't even time for a medical helicopter to pick him up, so a supply helicopter that was in the area had to be diverted to fly him out immediately. When he was loaded onto the helicopter, he assumed that it was a medical flight and that the crew chief was a flight medic. Jimmy started telling the crew chief what medical procedures he needed to save his life, the crew chief said, "Sorry pal, I'm just your ride." Jimmy's only advice at that point was, "fly fast." A stud of a man, Jimmy recovered from being 90% dead in a matter of months. He's still serving his country today as a firefighter outside of Cleveland.

Vic got hit by an RPG on a patrol. I was in the COC when I heard he had been hit. The company gunny told me that he hurt his leg so badly that they might have to cut it off and that he was not likely to survive. I checked the casualty tracker and saw that Vic was at Al-Asad, waiting for a flight to Landstuhl, Germany. As luck would have it, a mission came in to transport some detainees to Al-Asad later. I jumped on the mission, hoping to see Vic one last time. When I got there, I found a very beautiful, very alive Victor Lewis, who promised me he would not let anyone cut his leg off. Vic spent a lot of time in Bethesda recovering from his wounds. It was a long road. Vic struggled when he came home. He couldn't let go of the friends he lost and started drinking too much. He spent some time in prison. I guess in some ways, he never fully recovered.

To this day, Jimmy and Vic remain some of my most treasured friends. Jimmy Alunni-Finley and Victor Lewis are great men.

That picture. Looking at that picture of my friends. This whole crappy war and I was the last man standing in that picture, and barely standing at that.

There were some days that made me think that the remaining time I had on this earth was being measured in minutes. This was one of those days.

Our patrol had started out much like every other patrol. It was as hot as the face of the sun, no bad guys were anywhere in sight, and my shoulders were hurting after only the first hundred meters. The squad filed down one of the main north south routes along the outskirts of the city.

I was walking in the middle of the column behind a *jundi* when we passed the same burned-out car that we had passed eleven thousand times before. For whatever reason, the *jundi* looked inside the car as he passed.

I didn't know what the Arabic word for bomb was, but I am sure that the *jundi* was screaming it while he ran from the car and dove into a ditch. For a minute, I thought about running after him and doing the same thing, but my curiosity got the best of me. I poked my head in what had been the side window of the car and saw what had scared the hell out of the *jundi*. An 82mm mortar shell had been rigged with detonation cord and a triggering device, just waiting to wreck the patrol as we strolled by. Now, I thought, would be an appropriate time to run.

For some reason, either the trigger man was asleep, or he was waiting to hit the center of our patrol, the IED didn't explode. Good thing for the *jundi* and me, because if it had, we would have eaten it. The squad leader called in the bomb to the Combat Operations Center. They would dispatch the EOD team from the battalion to dispose of it. In the meantime, we would spread out in security positions, try not to get attacked, and wait. I offered to take the high ground on a nearby rooftop with another Marine. The elevated position afforded the best chance for spotting danger and was the farthest place away from the blast zone if the IED or a secondary IED detonated. Off to the roof we went.

The squad leader called for an EOD team from FOB Hit to come and render the device safe. "Rendering the device safe" was sort of misleading, because all that meant was blowing it up using a robot while no good guys were in the area. The Marine EOD team that was assigned to us was a phenomenal group of guys. They were on about their millionth tour, they were very professional, and they would stop at nothing to keep us safe. They didn't care if they had been out for three days straight with no sleep. If we called them and they would be there.

Unfortunately for us, the Marine team rotated home and was replaced by a Navy EOD team on their first tour. We waited about two hours for them to show up. When they did, they ran the standard EOD drill. A technician would remotely drive a bomb robot down to the IED. The robot was fitted with a camera so the technician could see what was going on from a safe distance and a mechanical hand for placing "counter charge" explosives, or explosives meant to safely detonate the bomb. The technician would then set a charge of plastic explosives in the proper place using the robot hand, drive the robot back, and blow the whole thing up. Even after you have seen it a hundred times, it was still cool. The entire car blew up in spectacular fashion and was engulfed in flames. I said a quick prayer of thanks for not being the one that was near it when it blew.

The other Marine and I walked down from our rooftop position and started walking back towards the squad leader to reform our patrol and continue our mission. We walked past the burning hulk of a car, fixated on the disaster that had been very narrowly averted. Just as I was abreast of the car, about ten meters away from it in about the same spot when I first saw the IED, the whole thing blew up again. A massive explosion came from the car and tossed me, ass over helmet, across the road.

I picked myself up out of the sand immediately and spit out some rocks that had managed to get in my mouth. For about a minute, I stood there frozen in a mixture of amazement and rage. My ears were ringing, the pressure in my gut made me want to throw up, and some of my teeth were loose. That blast stung. I looked at the car, then over disapprovingly at the EOD team, then at the car, then again at the EOD team.

What had happened was that the charge that the team set was not big enough to breach the door of the car and detonate the mortar round. The explosion we saw was the plastic explosive detonating and not the mortar round. The fire that resulted from the first explosion heated the mortar round until…well, until I walked by, and it decided to go off.

The petty officer in charge of the EOD team looked at me like he was staring at a ghost. A very angry, very sore ghost. He spoke first. "Wow…that's the first time that ever happened!"

I could have hit him I was so mad. "How 'bout that shit, hu?" I screamed as I marched towards him. I got about four steps before I read the expression on the petty officer's face. He fucked up and he knew it. He made a mistake that almost got me killed and I doubted very seriously if he would make that same mistake again. No sense in getting in a fist fight over it. I walked right past him to the squad leader, who was talking with our interpreter, and an Iraqi who had approached our patrol.

The interpreter passed on some very important and concerning news, "The guy said that the bad men have another bomb under the bridge up ahead." I looked at the Iraqi man, probably in his fifties but looking much older. He looked worn out. He had been watching people die for probably his whole life and he had enough of it. I had no doubt that he wanted the bomb gone more to protect his relatives who used the bridge for trade than to protect us, but it was nice to see someone with the courage to say something. I thanked the man, "*Shukran*."

I turned to the petty officer and informed him sharply, "Guess you get to try that again."

Our patrol accompanied the EOD team about three hundred

meters to the bridge, which was really nothing more than a four-foot-wide culvert over a sewage ditch. This provided a unique problem for the EOD team since they couldn't send the robot through the ditch to visualize the inside of the culvert. They would have to do it the old fashion way…by some poor schmuck sticking his head in it and looking around.

I had to fight back the urge to giggle as the petty officer came to the same realization that I had, that the robot was going to be useless. As mad as I was, I still didn't want to see him poke his head in a hole only to get it blown off by the trigger man who for sure would not miss this time.

The petty officer, it just so happened, also didn't want to get his head blown off. So, he took the easy way out. He knelt on the side of the bridge and took one of the quickest peeks into the hole that I have ever seen. If there had been a six-legged elephant in the culvert, he would not have been able to make it out, let alone an IED. Thinking that he had convinced the patrol that he had just done something worthwhile with his fake look, the petty officer stood up, threw his hands in the air and proclaimed for all to hear "It's good!" He then grabbed his partner, got into his Humvee, and started to leave.

I looked over at the patrol leader whose mouth was wide open at what he just saw. I started to speak with the exact same words he was thinking, "Corporal, we know there is a bomb under there and if we don't do something about it, it's gonna kill us. And if it misses us today, it will get 2nd Platoon tonight, or 3rd Platoon tomorrow. One way or another, one of our guys is eating that IED."

"What do you think then, Doc?"

I didn't say a word. I just walked over to the culvert, took off my helmet, got out my flashlight, knelt, and started crawling into the culvert. Without a word, the corporal followed suit taking off his helmet, kneeling beside me and placing his head right next to mine. I have to say that it probably wasn't the smartest thing in the world to have two guys in a position to get decapitated, but I appreciated the show of solidarity. We poked around for what seemed like a year but was probably about three seconds before someone spoke up.

"What the fuck are you doing?"

In unison, we both looked up to see who was doing the yelling. It was the petty officer.

"My *job*," I shot back, with about as much sarcasm as I could muster. I explained to the petty officer that someone was going to die if we didn't render that thing safe, so it was better to get it over with and not have to worry later.

Looking at me like I was crazy, he started shaking his head, and in a quiet, resigned tone, he told us to "go take some cover." We happily complied with his request and hid behind a wall as the petty officer prepared a plastic explosive charge to place into the culvert.

The call of "FIRE IN THE HOLE!" came three times before a giant mushroom cloud appeared in the sky. The massive blast wave made my teeth hurt as chunks of rock and dirt rained down for about a block in all directions. Had that IED gone off with Marines on top of it, there wouldn't have been enough pieces left to bury.

"Shit, that was big," I told the corporal. He just nodded in silence and said what I knew was coming. "Fucking Navy."

Our patrol had been out for eight hours, had barely walked a quarter of a mile, and I had almost been blown up three times.

I had such a distaste for EOD after that patrol that a week later, when we found our next IED, I decided to take matters into my own hands. I was on patrol with Sergeant Haney's squad from second platoon, when one of the SAW gunners spotted something placed next to a block wall, just below the surface. "IED!"

As soon as the warning sounded, Marines dove for cover out of the blast zone and quickly set up a defensive perimeter, scanning the area for an ambush, the trigger man, or a second device.

Sergeant Haney and I trotted quickly up to the SAW gunner to check out what he had found. Sergeant Haney was a half-crazy Marine who just transitioned from active duty. He had a thick New England accent and planned to be a hairdresser in Connecticut when he got home from Iraq. His claim to fame was getting us kicked out of six bars in a row in Las Vegas when we were on leave before deployment. Two of them before we even got in…fantastic guy and a great Marine, just odd.

I scanned the area closely with a telescopic rifle sight trying to make out the device. "I don't see any wires, maybe it's just UXO." Unexploded ordinance (UXO) littered the ground in Iraq. A bomb that was meant to go off (but didn't) can be very dangerous for a couple of reasons. First, it might just decide to go off with one of us next to it, and second, those rounds made perfect IEDs for insurgents.

"It ain't UXO, Doc," Haney explained. "They just buried the charge first and were going to come back and hook up the wires later."

Haney and I debated back and forth what exactly we were going to do with the charge. Clearly, we couldn't leave it here, or the next time a patrol walked by, they would be on the receiving end of it.

"Call EOD?" he asked me with a sort of a "do you have a better idea" tone in his voice.

We could call EOD, but it would take them hours to get there, all the while we would be sweating and just waiting to get attacked. We had been out all day and were low on water. I was hot, hungry, and didn't feel like turning a four-hour patrol into a month-long event. Plus, I was still salty about my last mission with EOD.

"It's not rigged to blow, I'll just blow it in place," I decided.

"How the hell are you gonna do that?" Haney asked.

"I'll just duct tape a frag to it, pull the pin, and jump behind the wall," I said, matter of fact. I had thought that frags, or fragmentation hand grenades, would be a dime a dozen in a war, but not really. I hadn't seen one used yet and I happened to be the only one in the company carrying one.

"Where the hell did you get a hand grenade from?" Haney asked, half smiling."I stole it."

"Fair enough. Everybody back up! Doc's gonna blow the IED in place." Sergeant Haney started to push the squad back to a safe standoff distance so they wouldn't be injured in the explosion. That was harder than it sounds, because everybody really wanted to see the insanity that was about to take place.

Blowing up a bomb with another bomb seemed simple on the surface of it, but as I worked through it in my mind, there were several very real opportunities for me to die. First, I had to walk up and inspect the IED. If I was wrong looking at it from a distance, and it was rigged to explode, then the trigger man that was watching me was sure to let me know I was wrong.

Next, I had to dig the explosive up to tape the frag to it. It was possible that the ordinance could detonate when I did that, or that the *mujahideen* could have booby-trapped the area, both of which would be bad things.

Lastly, I could fail to get far enough away from the blast and end up as one of the war's dumbest casualties.

As I was searching through my pack for my duct tape, I gave this whole plan a second thought. "Why is this a good idea?" I asked myself. The answer was simple. The longer we stayed in one spot the more certain we were to be attacked. And if we left the IED in place, we would be sure to find it again…when it blew us up.

I probed around the explosive to make sure that it was not set to explode, and very, very carefully, taped the fragmentation grenade to it, pin side up.

I squatted down, facing the retaining wall, ready to leap over it the second the pin was pulled. I tried to remember how many seconds delay the fuse on the grenade had, but I couldn't. I held both the charges into the sand with one hand and grabbed the pin with the other.

"Fire in the hole! Fire in the hole! Fire in the hole!" After the last announcement, I jerked the pin out sharply, and leapt over the wall, curling into a fetal position.

A massive explosion on the other side of the retaining wall, about three feet from me, was so loud I almost could not digest the noise. I could feel the blast wave move through my body almost in slow motion as I was tossed several feet along the ground.I picked my head up as the dirt and concrete was still falling around us. I shook the cobwebs out of my head and felt like I was going to throw up. I looked over to the other Marines to make sure everyone was okay, and locked eyes with Lance Corporal Shawn Beiger. Beiger was the smartest, fastest, and most cynical Marine I had ever met. He truly believed that he had died some time earlier, and every day in the Marine Corps was actually a day he was staying in purgatory.

Besides myself, Beiger had been the closest to the blast and was covered in dirt. He slowly and deliberately took off his glasses and picked a piece of earth from the corner of his eye. "How is it," he asked dryly, "that the guy who is supposed to save my life, almost killed me." Sorry, Shawn, not my best idea.

The war had started to break me. I was in the worst psychological state of my deployment. Someone couldn't slam a door without me thinking it was a mortar. And as soon as I realized that it was a door slamming and not a mortar, I would try and seek out and physically assault the person who did the slamming for making me think we were getting mortared.

Emotionally, I was done. All of the sticking my head in the culvert with the bomb in it, or blowing up the IED with the hand grenade, or running across the courtyard for a wounded Marine wasn't because I was brave, or stupid, or anything like that. It was because I had quit believing that I had any say in whether I lived or died. And after a while, I had just quit thinking that living was a possibility.

I had all but stopped talking to my wife. I would call her from the satellite phone, but we didn't have anything to say. Our lives had grown apart so much over the past year that I was more like a stranger than the husband that left Ohio.

The 2nd Marine Division had been pouring combat replacements into our battalion for weeks. Since the majority of our corpsmen had been killed, or wounded, fresh corpsmen straight from field med were streaming into the theatre. I remembered watching Vietnam War movies where no one ever wanted to befriend the new guy for fear the new guy would just get killed. I, for one, thought that was stupid, so I tried to make them feel at home the best I could.

I tried to be a friend to those…children, but I couldn't. I couldn't get past the fact that the only reason they were there was because Jeff and Joey and Vic and Travis and Jimmy…were gone. The mere fact that they were there made me resent them.

Not to mention that they were just so different than me. It wasn't that I felt that they hadn't earned my respect or friendship just because they hadn't seen combat. It was because I was jealous. When I looked at those kids…I saw kids. Eighteen, nineteen, twenty. When they talked, there was still joy in their voices; when they walked, they still held their heads high, when you looked into their eyes, there was still a soul behind them.

Most of our uniforms had been so destroyed that we had to replace them with fresh ones several times over. When we got a new blouse, it didn't have a name tape on it, it was just blank. Kind of symbolic. Like we weren't even persons anymore, just shells of a human beings walking around with a gun, waiting to kill or be killed.

These kids had sharp, crisp new uniforms. Uniforms with names. Names that represented people who still had their humanity. It was impossible to accept them, no matter how hard I tried.

Seeing them and realizing what I had become, and all I had lost, made me want to cry…except that I didn't want to use what few tears I had left on me, I wanted to save them for the next casualty.

CHAPTER 15
LEAVING IRAQ

How we (or at least me and the Marines of my company) left Iraq felt a little bit like we hadn't finished our job. We knew going in that our deployment would be about seven months of in country combat, but no one knew exactly when those seven months were going to end. A lot of people had figured out some approximate dates, but there was a range of about two or three weeks either way. So, when our tour wound down and fell within that window, there was a palpable change in the attitudes of the men of India Company.

Our patrol schedule got shorter. Our patrol routes were simpler. Our offensive missions were non-existent. Our contact with the *mujahideen* was as light as it could possibly be. Although no one would say it out loud, our battalion had been getting punched in the mouth for seven months, and no one wanted to be the last guy from 3/25 killed in Iraq. Accordingly, we had assumed a solidly defensive posture while we started the turnover with our replacement battalion.

On our patrols we would stay away from areas with which we knew contact was likely. Oftentimes we would just patrol to an area of town, occupy a vacant house, set up an observation post, and take turns sleeping. We had just about enough of war.

The only sure way I knew that my tour had ended was when my replacements started streaming through the gate on a convoy of seven-tons. Like a hundred and seventy-five camouflaged angles, it felt like we were being rescued. The 3rd Battalion First Marines (3/1) was finally moving in as our replacement.

Clean, fresh, and salty, the Marines of 3/1 were running after they jumped off the trucks. I was amazed at their energy and enthusiasm. I hadn't run except as a matter of self-preservation in months. Some days I barely had the energy to eat.

As happy as I was to see the 3/1 Marines, it would be fair to say that we didn't hit it off. They were an active duty battalion replacing a bunch of reservists. They were on their third tour; we had just finished our first. They were in the battle of Fallujah, the single biggest Marine operation of the war, we were not.

We had forty-eight Marines and corpsmen killed during our tour. This was the most men killed in one tour for any battalion, Army or Marine, in any conflict the United States had fought in since Beirut, Lebanon, twenty-two years before.

The 3/1 had been sent to our AO specifically because they were one of the Marine Corps' best battalions, and this was where the heaviest fighting was. Although no one in 3/1 ever said it, I was afraid that they thought we had lost so many men not because of our incredible operations tempo in a tough part of the country, but because we didn't know what we were doing.

That 3/1 thought we didn't know what we were doing was blatantly obvious. We tried to brief them on some of the problems we had encountered, particularly with civilian casualties. The insurgents had been probing us with unarmed civilians for the entire deployment. A patrol would be driving along, and for no good reason, a car would just refuse to stop. Despite flares, warning shots, and even attempts to disable the vehicle by shooting the engine, the car would just keep speeding towards the patrol.

The US had been in Iraq for three years now, and everyone with a car knew not to drive near the Americans, or they would think you were a car bomb and shoot you. So, that's what we did. Shoot the driver of the car that would not stop. Nearly all the time, the driver of the car was unarmed and not driving a car bomb.

So, the local population would be in an uproar over the shooting of an innocent man by the American infidels. Our command staff, not wanting to make enemies out of the local population (and rightly so), would then give us more restrictive rules of engagement, which meant we would shoot less drivers. This would result in vehicles being allowed to drive very close to our patrols. Once the insurgents saw that they could drive cars close to our patrols, one of them would, sooner or later, be a bomb. That bomb would wipe out more 3/25 Marines. Then the pendulum would swing back. The Marines, wanting desperately not to get blown up, would once again shoot any car that came close to them, killing more unarmed people.

Through all of this we had managed to achieve a sort of precarious balance. No one wanted to kill someone who didn't need killing, but no one wanted to get killed either. So, on every patrol, Marines agreed to accept a higher risk of personal harm when confronting potential threats. They would fire a few extra warning shots, shoot two flares instead of one, and wait until the last possible second just to make sure that they were not shooting someone who was innocent. Did this mean that they were exposing themselves to a greater likelihood of being killed or injured? Sure, it did. But it also gave the Marines the moral high ground of knowing they did all they could to act with honor and shoot only the people that really needed it.

We tried to explain this to the 3/1 Marines, but at least in my experience, they would have none of it. Fine by me. I wanted to pass on what I had learned not only because I was a professional, but because I felt I had a moral obligation to do so. If they didn't want to listen, that was okay. I was too whipped to argue.

A couple of months after we left, I was saddened to hear about what was labeled by the news media as the "Haditha Massacre." Some Marines from 3/1 had been killed in an ambush in Haditha, and in the resulting gunfire, there were some Iraqis killed who were unarmed.

The Marines were vilified; some were arrested and charged with murder. The news media naively exulted the plight of the Iraqi victim, claiming that those killed were purely innocent. The Marines claimed the opposite, that the only people killed were terrorists, plain and simple. All I am sure of is that what happened was infinitely more complicated.

It was a war, plain and simple. Marines were killed and their buddies in turn tried to return the favor to those responsible. The problem was that it was impossible to tell the good guys from the bad. Insurgents hid amongst civilians, fired from mosques, and paid innocent people to attack us. In repelling the attackers, those Marines had to decide in less than a second, who was good and who was bad. All the while their friends were dying. All the while they were praying to God to let them live.

My heart went out to those boys in 3/1. Countless times I had to make those same decisions on who needed to live and who needed to die. Which person to shoot, which not to shoot, which person to save, which not to save. Some of those choices have winners and losers, some only have losers.

Our departure came with little fanfare. One day, the First Sergeant came down and passed out the stick assignments: which squads would be leaving at what time. Over a span of about twelve hours, we packed up our gear, climbed into the back of trucks and slipped quietly into the night.

I passed on our excess medical equipment to the chief corpsman from 3/1 and explained to him quickly some of the problems we were having.

"They stopped pumping out the porta shitters about a week ago, so they are beyond full. You'll need to fix that. Here's a box of extra tourniquets, we have been using the hell out of them. The *jundis* that have USMC med kits are the Iraqi medics who are worth a shit and can speak a bit of English."

Next, I broke the bad news to him. "Our guys bought the TV, couch, and rice cooker out in town. There are some air wing guys back at Al-Asad who want to buy them, but I wanted to give you first dibs." Hocking all the stuff one unit pillaged to the guys who replaced them was a time-honored tradition. It happened to us when we arrived, it happened to the guys before us, and it continued to happen long after we left.

"How much are you askin?"

"A hundred bucks."

"Deal," the 3/1 corpsman collected $100 from some of his Marines and paid me as I was climbing into the back of the seven-ton for my last trip out.

For the first time in forever, I cracked a smile as I rode off. We hadn't paid a dime for that stuff; it was all stolen. Not to mention that I didn't have a buyer for it and was going to have to leave it for him anyway, cause I sure as heck couldn't fit it into the back of the truck.

I don't know what made me happier, scamming those arrogant 3/1 pricks out of $100, or leaving this whole mess of a war.

I rode from the FOB for the last time in the back of that truck, packed next to sixteen Marines who sat silent and motionless. Months of emotional torture had left me too void to even process what had finally become a reality for me.

I had made it. I lived.

We loaded off the trucks at Al-Asad to await our flight home. Hours later, I was standing on the tarmac, waiting to board a C-130 headed for Kuwait. As the ramp on the huge plane dropped, a platoon of fresh Marines filed off.

The whole scene from my first day in Iraq was now playing back, in reverse. I wasn't the brand new kid running off the plane anymore, scared and confused. I was the man standing in line to go home, sullen and beaten.

When I first arrived, I wondered why the 1/7 Marines weren't happy to leave. Now I know it's because no one is happy when they leave a war. I left a place I bled to protect. I left friends who will never come home. I left the person I used to be. I was thankful to be alive, but I was certainly not happy to be leaving. In fact, leaving Iraq was one of the saddest days of my life.

After Mother's Day, I was constantly searching for a way to fill the void of loss, but I couldn't find one. And after each new loss, that void just kept getting bigger and bigger. I thought that I would feel better after I saw David. Meeting him helped, but it was not a cure. I thought I would feel better after I saw Maria. We have visited several times since my return and talk regularly. Each time I talk to her, I feel a little more whole, and a little sadder all at the same time.

Louise, the kids, and I visited Maria at her home in Long Island. We visited Jeff's grave in Calverton National Cemetery. Appropriately, Jeff is buried next to Navy Seal Lieutenant Michael Murphy who was killed in Afghanistan and was awarded the Medal of Honor. I can't think of a more appropriate place for a Medal of Honor recipient to be buried than next to my friend, Jeff Wiener.

We came back to Maria's house after the visit to the cemetery and cooked dinner. Maria had me do the grilling and I watched as our children swam in the pool. As I stood there, cooking on Jeff's grill, with Jeff's wife, playing with Jeff's kids, I started to cry. Louise immediately came over to comfort me, knowing that I was moments away from a complete breakdown.

"It should be him! He should be here cooking for his family, not me! It's NOT FAIR!" Although I never felt like I should have been the one to die instead of Jeff, I always felt like he at least deserved the luck that I had. He had earned that through his life. He was the better man.

Louise held me tightly as she tried to hug my pain away. "I know honey, it's not fair at all."

I have since learned that fairness has nothing to do with it. And that I drove myself mad trying to figure out what's fair and what's not. Things aren't fair, they just are.

That night I would tell Maria about the day Jeff died, what he said to me before he left on his last patrol, and how it was his wish that she remarry. It was difficult to stay composed, but I felt good for having a purpose.

Foolishly, I thought that there would be one thing that would certainly bring me closure. The death of Abu Musab al-Zarqawi. Zarqawi was the leader of Al-Qaeda in Iraq, and he operated almost exclusively in our area. He was directly responsible for the attack that killed Jeff, and almost killed me. He was also responsible for the ambush of six of our snipers who were killed, one of them who was captured and defiled on television. Zarqawi had the blood of forty-eight of my friends on his hands, along with that of countless other Americans. The United States Marine Corps had been trying to kill him for quite a while.

On June 7th, 2006, the US Air Force finished the job.

I called Maria immediately after hearing the news on CNN. "They got him. They got the guy that killed Jeff."

She simply said one word: "Good."

We hung up. I took a deep breath and realized that I did not feel one bit different after Zarqawi was killed than before. One thing that doesn't ease the pain of death, is more death. Healing is a process that comes in small steps, over time, and is not helped along by more dead people, no matter how bad they are.

My uncle Andy was awarded a Silver Star as an Army Ranger in Vietnam. During a visit to my home, Andy asked me a seemingly simple question.

"What do you think about going to Iraq?"

I answered that question in an altogether different way when someone of his stature asked it than I did when a fourth-grade teacher asked it while I was giving a Veteran's Day speech to her class.

"Andy, after all the terrible things I saw, and did, and after everything that happened to me and my friends, if I had the same decision to make again, I would go in a minute. I don't know why, I just would. I couldn't live with myself if I didn't."

Andy nodded his head in agreement and told me a story that probably sums up better than I ever could why I would make that same decision again.

"Taylor, I've met a lot of very successful people in my life. And when they find out I was in Vietnam, and was wounded, the vast majority of them say the same thing. They regret not having gone there themselves. For all their achievements in business, for all the millions of dollars they have made, and all their contributions to society, they still regret not having gone to Vietnam. You and me, we don't have to worry about that."

I went through quite the moral dilemma concerning whether to enlist or not. While my choice, and hopefully even my motivation, is obvious now, I don't want to give the impression that any type of finality came with my decision. I had hoped that by making my decision I would get "inner peace" or, at the very least, an end to the incessant conflict of emotions that was created by having to choose between my family and my beliefs.

I didn't get that.

I would venture to say that I am even more conflicted now. My wife, although consummately supportive, still thinks I made the wrong decision, and she has a good body of evidence to support her claim. I put her through a year of intense psychological torment. I missed the birth of my son. I gave my daughter some yet-to-fully-be-discovered abandonment issues. I lost my best friend. I was wounded twice physically, and probably permanently scarred emotionally. I made a choice that changed who I am as a person, that changed how people see me; and that changed the relationship within my family.

Despite all of that, the time I spent in Iraq was spent with the greatest human beings the world has ever produced. I would never give one second of that time back and I would give anything to have just one minute more with them.

Everything happens for a reason. To Jeff, to all of us. Jeff telling me he wanted Maria to remarry, seeing Josh for the last time, telling me in the hallway an hour before he died that it would "be alright," and him spending one last magical week with his wife in Vegas. It was all for a reason.

I don't know if Jeff instinctively knew what was going to happen to him, or whether it was just his nature to make every decision count, and to live every day like it was his last. With passion. With purpose.

I do know that I learned a lot about how to live my life from him, and I am forever grateful for the lesson.

So, that is how I try to celebrate his memory, and the memories of the other forty-seven Marines I left overseas. I try to be the best father, son, and American I can be. I vote, volunteer, go to church, raise wonderful children, eat right, speak out against injustice, champion causes passionately, cherish the sunsets over Lake Erie, and help those in desperate need. I try to be the best I can be, in everything I do, every day.

This is how I have tried to live, every day since I have returned. And I pray that when I have come to the end of my time on this earth, I can look Jeff, and all the other friends I lost in the eyes and feel like I have earned the privilege of their sacrifice.

Viktor Frankel, when he was writing about his time in the Nazi death camp Auschwitz, said, "The best of us, did not survive." There are many men that I have written about in this book, still living, that are better men than I can ever hope to be. But the forty-eight brothers that did not come home, and the ones who have died tragically since, were truly the best of 3/25.

ABOUT THE AUTHOR

Taylor Cleveland was born and raised in Conneaut, Ohio, a small lakeshore community outside of Cleveland which was founded by his nine-times great-grandfather in 1796. He has worked as a police officer, firefighter, and paramedic in his local community. He currently works as a police detective and lives in Conneaut with his wife of 23 years, Louise. Together, they have three children.

After September 11th, 2001, Taylor entered the US Military and volunteered as a United States Navy Reserves Corpsman. He deployed to the Al-Anbar Province of Iraq in 2005 as the senior line company corpsman for India Company, 3rd Battalion, 25th Marine Regiment.

Taylor is an avid supporter of Feed Our Vets, a non-profit organization working to feed Veterans in need in Ashtabula County, Ohio. All of the proceeds from *Mother's Day* will go to Feed Our Vets to support their mission.

ABOUT THE PUBLISHER

TACTICAL 16

TACTICAL16 PUBLISHING

Tactical 16 Publishing is an unconventional publisher that understands the therapeutic value inherent in writing. We help veterans, first responders, and their families and friends to tell their stories using their words.

We are on a mission to capture the history of America's heroes: stories about sacrifices during chaos, humor amid tragedy, and victories learned from experiences not readily recreated—*real stories from real people.*

Tactical16 has published books in leadership, business, fiction, and children's genres. We produce all types of works, from self-help to memoirs that preserve unique stories not yet told.

You don't have to be a polished author to join our ranks. If you can write with passion and be unapologetic, we want to talk. Go to Tactical16.com to contact us and to learn more.

All of Tactical 16's books are available on our online bookstore, T16Books.com. Visit it today to see more books from our selection of authors and to find a new adventure to read!

T16BOOKS

Printed in the USA
CPSIA information can be obtained
at www.ICGtesting.com
JSHW072244140823
46471JS00005B/18

9 781943 226832